Conference Proceedings of the Society for Experimental Mechanics Series

Series Editor
Kristin B. Zimmerman, Society for Experimental Mechanics, Inc., Bethel, CT, USA

D1824970

The Conference Proceedings of the Society for Experimental Mechanics Series presents early findings and case studies from a wide range of fundamental and applied work across the broad range of fields that comprise Experimental Mechanics. Series volumes follow the principle tracks or focus topics featured in each of the Society's two annual conferences: IMAC, A Conference and Exposition on Structural Dynamics, and the Society's Annual Conference & Exposition and will address critical areas of interest to researchers and design engineers working in all areas of Structural Dynamics, Solid Mechanics and Materials Research.

More information about this series at http://www.springer.com/series/8922

Sharlotte L. B. Kramer • Rachael Tighe • Ming-Tzer Lin • Cosme Furlong
Chi-Hung Hwang
Editors

Thermomechanics & Infrared Imaging, Inverse Problem Methodologies, Mechanics of Additive & Advanced Manufactured Materials, and Advancements in Optical Methods & Digital Image Correlation, Volume 4

Proceedings of the 2021 Annual Conference on Experimental and Applied Mechanics

 Springer

Editors
Sharlotte L. B. Kramer
Sandia National Laboratories
Albuquerque, NM, USA

Ming-Tzer Lin
National Chung Hsing University
Taichung, Taiwan

Chi-Hung Hwang
National Applied Research Laboratories
Taiwan Instrument Technology Institute
Hsinchu, Taiwan

Rachael Tighe
School of Engineering
University of Waikato
Hamilton, New Zealand

Cosme Furlong
Mechanical Engineering Department
Worcester Polytechnic Institute
Worcester, MA, USA

ISSN 2191-5644   ISSN 2191-5652   (electronic)
Conference Proceedings of the Society for Experimental Mechanics Series
ISBN 978-3-030-86747-8   ISBN 978-3-030-86745-4   (eBook)
https://doi.org/10.1007/978-3-030-86745-4

This Springer imprint is published by the registered company Springer Nature Switzerland AG
The registered company address is: Gewerbestrasse 11, 6330 Cham, Switzerland

Preface

Thermomechanics and Infrared Imaging, Inverse Problem Methodologies, Mechanics of Additive and Advanced Manufactured Materials, and Advancement of Optical Methods and Digital Image Correlation represents one of four volumes of technical papers to be presented at the 2021 SEM Annual Conference and Exposition on Experimental and Applied Mechanics organized by the Society for Experimental Mechanics scheduled to be held during June 14–17, 2021. The complete proceedings also include volumes on: *Dynamic Behavior of Materials; Challenges in Mechanics of Time-Dependent Materials, Mechanics of Biological Systems and Materials, and Micro- and Nanomechanics; and the Mechanics of Composite, Hybrid and Multifunctional Materials, and Fracture, Fatigue, Failure and Damage Evolution.*

Each collection presents early findings from experimental and computational investigations on an important area within experimental mechanics; residual stress, thermomechanics, and infrared imaging inverse problem methodologies; and the mechanics of additive and advanced manufactured materials being a few of these areas.

In recent years, the applications of infrared imaging techniques to the mechanics of materials and structures have grown considerably. The expansion is marked by the increased spatial and temporal resolution of the infrared detectors, faster processing times, much greater temperature resolution, and specific image processing. The improved sensitivity and more reliable temperature calibrations of the devices have meant that more accurate data can be obtained than were previously available.

Advances in inverse identification have been coupled with optical methods that provide surface deformation measurements and volumetric measurements of materials. In particular, inverse methodology was developed to more fully use the dense spatial data provided by optical methods to identify mechanical constitutive parameters of materials. Since its beginnings during the 1980s, creativity in inverse methods has led to applications in a wide range of materials, with many different constitutive relationships, across material heterogeneous interfaces. Complex test fixtures have been implemented to produce the necessary strain fields for identification. Force reconstruction has been developed for high strain rate testing. As developments in optical methods improve for both very large and very small length scales, applications of inverse identification have expanded to include geological and atomistic events. Researchers have used in situ 3D imaging to examine microscale expansion and contraction and used inverse methodologies to quantify constitutive property changes in biological materials.

Mechanics of additive and advanced manufactured materials is an emerging area due to the unprecedented design and manufacturing possibilities offered by new and evolving advanced manufacturing processes and the rich mechanics issues that emerge. Technical interest within the society spans several other SEM technical divisions such as composites, hybrids and multifunctional materials, dynamic behavior of materials, fracture and fatigue, residual stress, time-dependent materials, and the research committee.

The topic of mechanics of additive and advanced manufacturing included in this volume covers design, optimization, experiments, computations, and materials for advanced manufacturing processes (3D printing, micro- and nano-manufacturing, powder bed fusion, directed energy deposition, etc.) with particular focus on mechanics aspects (e.g., mechanical properties, residual stress, deformation, failure, rate-dependent mechanical behavior).

With the advancement in imaging instrumentation, lighting resources, computational power, and data storage, optical methods have gained wide applications across the experimental mechanics society during the past decades. These methods have been applied for measurements over a wide range of spatial domain and temporal resolution. Optical methods have utilized a full range of wavelengths from X-ray to visible lights and infrared. They have been developed not only to make two-dimensional and three-dimensional deformation measurements on the surface, but also to make volumetric measurements throughout the interior of a material body.

The area of Digital Image Correlation has been an integral track within the SEM Annual Conference spearheaded by Professor Michael Sutton from the University of South Carolina. The contributed papers within this section of the volume span technical aspects of DIC.

The conference organizers thank the authors, presenters, and session chairs for their participation, support, and contribution to this very exciting area of experimental mechanics.

Hamilton, New Zealand   Rachael Tighe
Albuquerque, NM, USA   Sharlotte L. B. Kramer
Taichung, Taiwan   Ming-Tzer Lin
Worcester, MA, USA   Cosme Furlong
Hsinchu, Taiwan   Chi-Hung Hwang

Contents

Chapter 1
Super-Resolution Optical Microscopy to Detect Viruses (SARS-CoV-2) in Real Time

C. A. Sciammarella, L. Lamberti, and F. M. Sciammarella

Abstract This article describes the use of evanescent light fields to directly observe and detect the newly discovered coronavirus SARS-CoV-2 that causes COVID-19. The proposed technique provides a low-cost, fast, and highly accurate method of detection. This approach builds from previous work from the authors that enables microscopic observations of nano-objects with the accuracy of nanometers and sensitivities of the order of fraction of a nanometer.

Keywords Optical super-resolution · Evanescent fields · Holography at the nanoscale · Carrier gratings · Digital microscopy · Covid-19 virus · Techniques of detection of covid-19

1.1 Introduction

The observation of objects in the nanometric range is at the cutting edge of current technology and is of paramount importance in the detection of viruses such as the SARS-CoV-2 which causes COVID-19. Currently the virus images are observed with electron microscopes that require freezing the virus at extremely low temperatures. This is a time-consuming and costly process. To effectively combat this virus, it would be critical to make observations of live samples in real time. The authors present previous work that can be extended to the observation of the COVID-19 virus at room temperature using visible light from an optical microscope in real time. In 2009, the authors [1] presented a comprehensive approach using an optical microscope with a methodology that obtained metrological information from prismatic crystals and spheres in the ranges of 10–200 nm that achieved a level of resolution of 0.1 nm, with the accuracy of ±3 nm comparing experimental results with known theoretical sizes.

To achieve the resolution and accuracies with this setup the following conditions must be met. First, it is important to establish an evanescent field within the field of view, via laser light by total internal reflection. The laser light has a similar function to feeding illumination in a laser resonator. It excites the objects to generate light that is defined by the molecular structure of the observed object composition, geometry, and dimensions. The second condition is having the presence of gratings in the optical system. These gratings provide carriers encoding the sought information. The third condition is having a small ball-lens in the sub-millimeter range. This lens acts as a relay in the microscope setup that transmits the near-field information to the far field. At the same time provides the geometry required to produce a nano-resonator in the electromagnetic field. Finally, signal decoding software extracts information from the images that allows reconstruction of the observed objects. Since 2005, the authors made use of evanescent fields in several technical applications [2–7] and in larger fields of view, utilizing several optical configurations. Based on this experience, the authors propose the design of a super-resolution digital microscope that would be capable of observing COVID-19 in real time at an affordable cost. An important characteristic of the system is the utilization of artificial intelligence software to make real-time detection feasible. The following sections will summarize the main topics necessary to set the foundations of the proposed technology.

C. A. Sciammarella (✉)
Department of Mechanical, Materials and Aerospace Engineering, Illinois Institute of Technology, Chicago, IL, USA
e-mail: sciammarella@iit.edu

L. Lamberti
Dipartimento Meccanica, Matematica e Management, Politecnico di Bari, Bari, Italy

F. M. Sciammarella
MxD, Chicago, IL, USA

© The Society for Experimental Mechanics, Inc. 2022
S. L. B. Kramer et al. (eds.), *Thermomechanics & Infrared Imaging, Inverse Problem Methodologies, Mechanics of Additive & Advanced Manufactured Materials, and Advancements in Optical Methods & Digital Image Correlation, Volume 4*, Conference Proceedings of the Society for Experimental Mechanics Series, https://doi.org/10.1007/978-3-030-86745-4_1

1.2   Background on the Generation of Optical Signals at the Near Field

When analyzing super-resolution images to obtain both shape and size, it is necessary to understand the processes of generation of the signals that contain information about the properties and dimensions of the objects. There are two basic ranges that depend on the relationship of the illuminating beam wavelength λ to the dimensions of the observed objects. The dimensions of the object are such that minimum dimension dmin> λ, and the other case is when the dimensions of the object are smaller than the wavelength of the object, dmin<λ. In the case of objects smaller than the wavelength of light let us consider a simple case, observations of prismatic shapes. Experimental evidence shows that observing a face of a prism, for example, the face with dimensions [Lo, Wo], Fig. 1.1, one gets 3 shifted images (Fig. 1.2) corresponding to three partially overlapping wavefronts. In [1] a model was introduced for the signal generation by objects smaller than the wavelength of light, nanocrystals, nanospheres. The model is based in the Brillouin scattering process in the ultrasound range [8, 9]. Scattering in this context means collision between particles, in this case collision between phonons and electrons. The Brillouin scattering in the ultrasound range is based in the piezo-electric effect in materials. Piezoelectric effect is the ability of certain materials to generate electric charges in response to applied mechanical stresses. An acoustic wave is generated in a crystal that propagates with the velocity of sound in the material medium. This acoustic wave produces the conversion of mechanical energy of the deformed into the electro-magnetic energy of emitted light.

  The process of getting the crystal dimensions gives an insight on the different steps that lead to the determination of the prism proportions. Dimensional information is obtained through the eigen mode of vibration of the prism. The prism vibrates axially along the symmetry axis, alternatively expanding and contracting. The phenomenon that produces the observed shifted images of a wavefront is called Bragg's diffraction, in analogy to X-ray diffraction where a similar process takes place. The acousto-optic effect is akin to the photoelastic effect, in the sense that the permittivity ε of an observed crystal is changed due to strain components ε_{ij}. The equivalent of a standing diffraction grating is formed by the sound waves propagating in the medium. Through this process mechanical energy is transmitted to the electronic field. This diffraction pattern is characterized by a diffraction angle θ_n with respect to the direction of propagation of the light wavefront, and is given by,

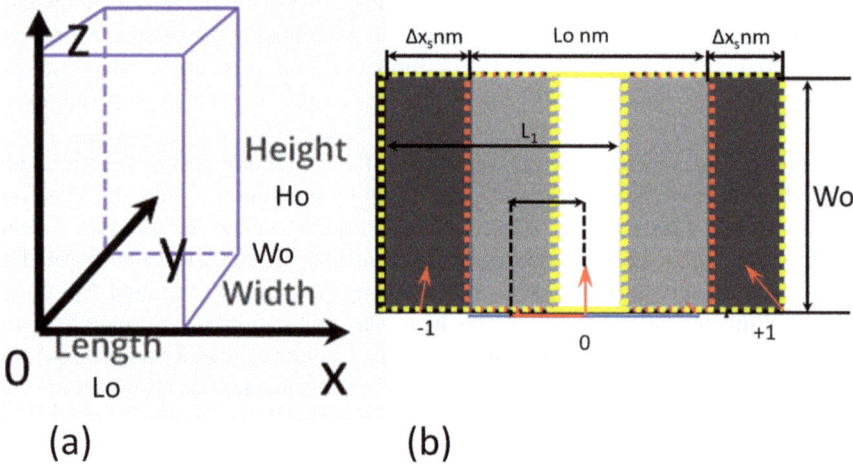

Fig. 1.1 (**a**) Rectangular prismatic single crystal; (**b**) Observed face Lo X Wo of a prism, three shifted images of the face are observed

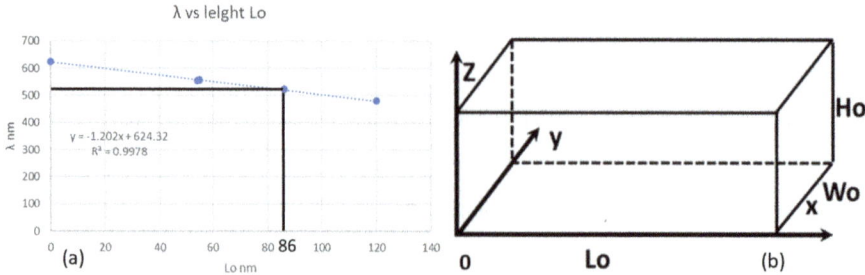

Fig. 1.2 (**a**) Wavelength λ of the resonant modes as a function of the crystals length Lo; (**b**) Observed single crystals geometry and corresponding reference axis

$$\sin\theta_n = n\frac{\lambda}{2\Lambda} \qquad (1.1)$$

In Eq. (1.1) λ is the wavelength of the optical wave, Λ is the wavelength of the acoustic wave and n is the diffraction order of the generated light. Due to energy considerations most of the energy concentrates in two diffraction orders. The resonance modes in the electromagnetic field of the observed crystals produce the emission of wavelengths that depend on two fundamental factors, the chemical composition of the crystal and on the dimensions of the crystal. This model predicts that the wavelength of light generated by a given crystal is a function of the length of the crystal Lo. The Lo designation is given taking into consideration the position of the prism with respect to the illuminating laser beam and the excited eigen mode in the prism. This prediction is confirmed by the graph of Fig. 1.2 where experimental data of the wavelengths determined for different nanocrystals of sodium chloride are plotted together with the values predicted by the model of [1].

Figure 1.3 is an additional experimental proof that the light emitted by a crystal as predicted by the developed model is a function of the crystal length. Figure 1.3a is a gray scale image of one of the observed crystals, Lo = 86 nm. Figure 1.3b is a color image of the same crystal. Although the geometrical scales of both images are not the same, visually it is possible to see that both images correspond to the same crystal. As mentioned before, the illuminating laser wavelength is $\lambda = 0.6328\ \mu m$, the wavelength of the observed image is, Fig. 1.4, 524 nm that corresponds to a light green hue. Figure 1.4 shows that it is possible to get the prism dimensions Lo and Wo from the images captured within the setup. The other unknown dimension of an observed prism is Ho, the height of the prism.

Figure 1.4a shows the enlarged image of a nanocrystal. Figure 1.4b shows the atomic reconstruction of the crystal from theoretical considerations based on monomers of the crystal, also showing the corresponding reference Cartesian-axes. Figure 1.4c shows the sizes Lo and Wo of the crystal obtained by utilizing edge detection algorithms and shows the amount of the image shift. Two of the 3 unknows are determines as it is shown in this example. It is now necessary to obtain the magnitude of Ho, the third unknown, the depth of the nanocrystal. There are two techniques that are possible to apply [1]. One is based on the magnitude of the shift Δx_s, Fig. 1.3c. We will explain the second technique that is based on the principles of the observation of phase objects in holographic interferometry.

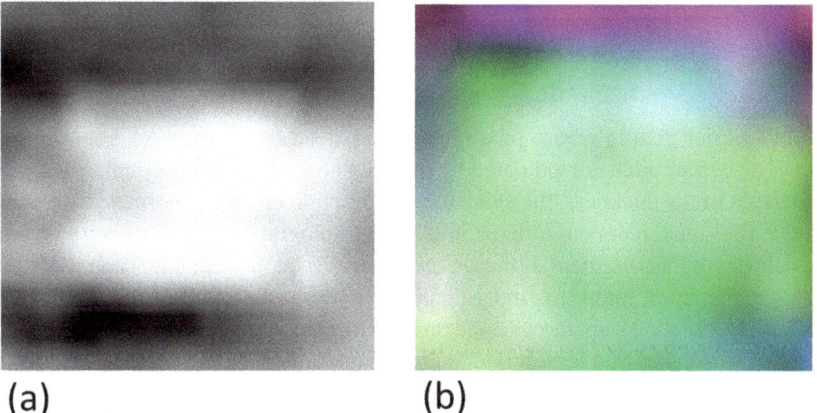

(a)   (b)

Fig. 1.3 NaCl nanocrystal of length 86 nm: (**a**) grayscale image; (**b**) image of the crystal captured by a color camera

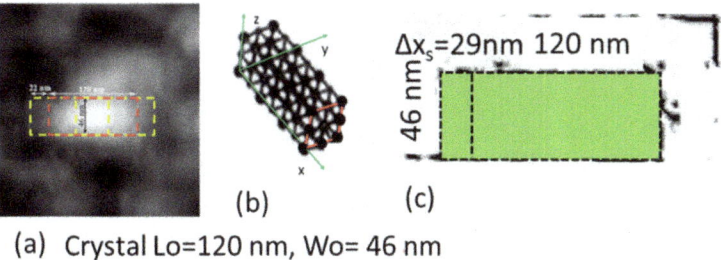

(b)   (c)

(a) Crystal Lo=120 nm, Wo= 46 nm

Fig. 1.4 (**a**) Enlargement of area surrounding the crystal; (**b**) Atomic model of the nano-monomer of the sodium chloride; (**c**) contour of the crystal and of the shifted image obtained using edge detection

In the second technique, technique, we make use of the different diffraction orders of a grating and it is a major component of the optical setup utilized to obtain the images of the nanocrystals. Let us consider the quasi-monochromatic coherent scalar wave emitted by a nano-sized crystal. The actual formation of the image of the nanocrystal is similar to a typical lens hologram of a transparent object illuminated by a grating. The gratings are carrier gratings that can be utilized to extract optical path changes. This type of setup to observe phase objects has been used in phase hologram interferometry as a variant of the original setups [10, 11]. When the index of refraction in the medium is constant, the rays going through the object are straight lines. The prismatic object emits a beam normal to its faces; the optical path s_{op} through the prism is given by the integral,

$$s_{op}(x, y) = \int_0^t n_i(x, y,) \, dz \tag{1.2}$$

In Eq. (1.2) the direction of propagation of the beam is the z-coordinate and the analyzed plane wavefront is the plane x–y; $n_i(x,y,z)$ is the index of refraction of the medium.

The change experienced by the optical path is given by,

$$\delta_{op}(x, y) = \int_0^t \left[n_i(x, y, z) - n_o \right] dz \tag{1.3}$$

In Eq. (1.3) t is the thickness of the medium (called Ho in Fig. 1.1). Assuming that:

$$n_i(x, y, z) = n_c \tag{1.4}$$

In Eq. (1.4) n_c is the index of refraction of the observed nanocrystals, Eq. (1.4) then becomes,

$$\delta_{op}(x, y) = \int_0^t \left[n_c(x, y, z) - n_o \right] dz = (n_c - n_o)t \tag{1.5}$$

By transforming Eq. (1.5) into phase differences and making $n_o = n_s$, where n_s is the index of refraction of the saline solution containing the nanocrystals, one can write:

$$\Delta\Phi = \frac{2\pi}{p}(n_c - n_s)t \tag{1.6}$$

In (1.6) p is the pitch of carrier grating whose phase has experienced the change $\Delta\phi$ going through the thickness of the nanocrystal Ho, Fig. 1.1. The carrier pitch is modulated by the thickness of the nanocrystal. For a given diffraction order the optical path is a linear function of the thickness, the evanescent diffraction order 0 is normal of the surface of the nanocrystals. The conclusion is then that the same treatment utilized in classical holographic interferometry of transparent objects can be utilized for the nanocrystals depth measurement. For each analyzed crystal, a particular pitch p of the carrier is selected from the FT of the image. The frequency of interest is individualized in the background of the observed object and a diffraction order is selected and filtered. In a second step, the selected order is located in the FT of the observed object. A filter size is then adopted to pass additional of harmonics that correspond to the modulation of the carrier produced by the thickness Ho. Those additional frequencies carry the information of the change in phase produced by the change of optical path. In the next step, the phase of the modulated carrier and the phase of the unmodulated carrier are computed. The change of phase is introduced in Eq. (1.6) and the value of Ho is computed. The phase difference may not be constant throughout the prismatic crystal face

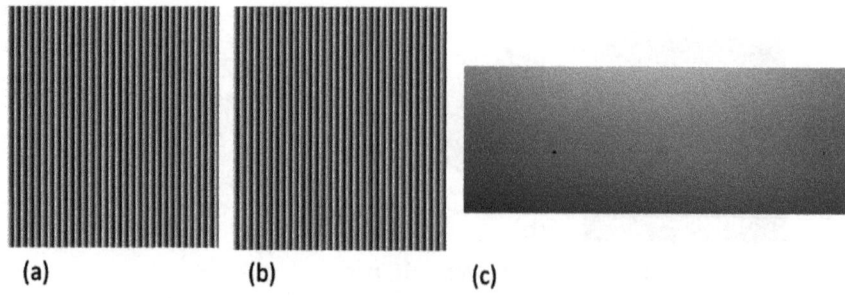

Fig. 1.5 Phase determination for the nanocrystal of length L = 120 nm: (**a**) reference phase of the carrier fringes; (**b**) phase of carrier fringes modulated by the nanocrystal; (**c**) phase difference in the region of the nanocrystal represented in levels of gray

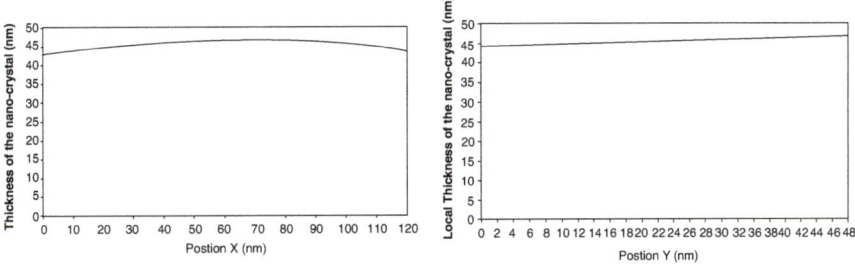

Fig. 1.6 (**a**) Cross section of the crystal Lo = 120 nm along the x-axis; (**b**) Cross section of the crystal Lo = 120 nm along the y-axis

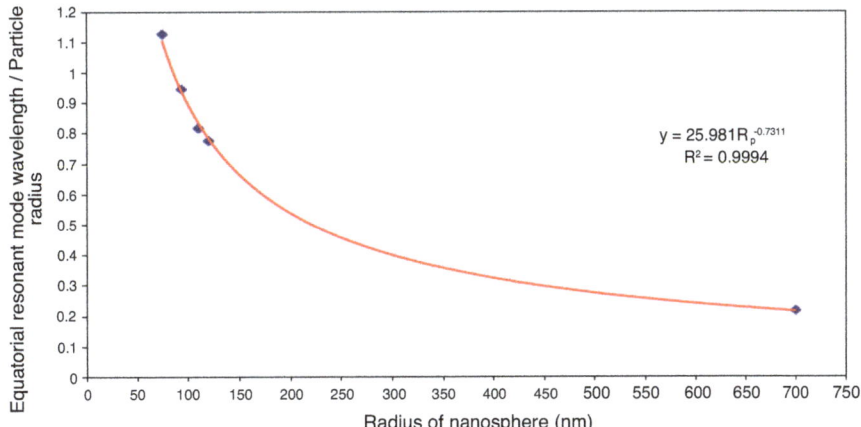

Fig. 1.7 Correlation of the nanospheres radius and the equatorial wavelength expressed as a fraction of the radius

since it is unlikely that the crystal face is parallel to the image plane of camera. Therefore, an average thickness is computed. Hence, depth can be computed as an average of the depth coordinates of points of the face. Figure 1.5 illustrates, for the crystal of Lo = 120 nm, the different steps of the process described above. The reference phase pattern corresponds to a carrier in a region near but not in the correspondence with the crystal image. It is acquired by applying in the FT of the image a 1×1 filter (Fig. 1.5a) corresponding to the pitch p = 5.53 nm. The carrier modulated by the thickness of the crystal, Fig. 1.6b, is obtained by selecting the FT of the image of the same order selected in the background but with a filter that allows additional orders to pass. Figure 1.5c represents the change of phase in gray levels. Figure 1.6a represents the cross section of the crystal along the longitudinal direction, coordinate x. Figure 1.6b represents a transversal cross section in the y-direction. The value of the depth resulting from averaging the phase plotted in Fig. 1.5c is 46 nm.

1.3 Observation of Nanospheres

The same principle that applies to the formation of images that was presented in Sect. 1.2 for nano prismatic crystals is valid also in the case of nanospheres; the observed pattern of resonance is a function of the length of the perimeter of the equatorial circle of the excited form of vibration of the nanosphere. Micro/nanospheres made of transparent dielectric media are excellent optical resonators. Both theoretical and experimental studies on the resonant modes of micro/nanospheres are available in the literature. Of particular interest are the modes localized on the surface of the sphere, along the equator. These modes are called *whispering gallery modes* (WGM). WGM result from light confinement due to total internal reflection inside a high index spherical surface immersed in a lower index medium. Of all resonant geometries a sphere has the capability of storing and confining energy in a small volume. Figure 1.7 shows the correlation between the nanospheres radius and the equatorial wavelength expressed as a fraction of the corresponding radius. It is interesting to point out that the point corresponding to r = 700 nm corresponds to a numerical solution of a polystyrene nanosphere [12]. There is an excellent agreement between the experimental values the authors obtained and the numerical solution, $R^2 = 0.9994$.

1.3.1   Optical Setup

The optical system consists of a CCD camera attached to a digital microscope. Figure 1.8 represents the optical arrangement. A glass sphere is supported on a microscope slide; this sphere has a dual role. The sphere acts as a ball lens in the microscopic system, at the same time as it will explained later, contributes to the generation of the electromagnetic field that produces the images of the nano-objects. The optimal numerical aperture of the object space of the spherical lens is limited by the spherical aberrations. The spherical aberration limits the maximum achievable resolution in the image. The optimal theoretical position of the ball lens pupil is in the center of the ball lens. Since such a configuration is difficult to implement, the next practically acceptable position for the aperture stop is directly behind the ball lens, in this case the surface of the supporting glass slide. The focal distance f of this lens can be computed using the equations of geometric optics. The source of the illumination generating the image of the nanospheres and the carrier gratings covering the image of the nanospheres as we have seen are the resonant electro-magnetic oscillations generated in the observed objects by the laser beam. The whole region of contact between the observed objects and the microscope slide is the equivalent of a micro-laser resonator. The light is generated in the nano-objects and focused on the focal plane of the microlens. The digital microscope shown in Fig. 1.8 is focused to the plane of best contrast of the image, which is close to the ball lens focal plane. The image is not formed by plane wave fronts illuminating the nano-objects, as is the case with the classical Mie's solution for a diffracting objects. In the present case, a large number of evanescent waves impinge in the nano-objects and then generate propagating wave fronts. The wavefronts emitted by the nano-objects have the structure of pseudo non-diffracting propagating waves. The light emission by the nanospheres is produced by multi-polar resonance of the nano-objects caused by the electromagnetic field generated by the evanescent waves. Evanescent beams are converted into actually propagating beams when they interact with the nano-objects. A complete analysis requires a vectorial solution of the Maxwell equations; as a first approach a scalar form can be utilized. The observed light fields can be considered as solutions of the Helmholtz equation.

$$\left(\nabla^2 + k^2(r)\right)W(r,\theta) = 0 \tag{1.7}$$

In Eq. (1.7) (r,θ) denotes the transverse coordinates, ∇^2 is the Laplacian, $k(r)$ is the wave vector, $W(r,\theta)$ is the scalar electromagnetic field. These waves produce images in the camera attached to the microscope. The pseudo non-diffracting wavefronts can travel long distances or go through an optical system without the diffraction changes experienced by ordinary wavefronts. The validity of this approach is supported in the following references [13–16].

Figure 1.8 shows a detailed view that graphically illustrates the process of formation of gratings that produce the lens-holograms of the observed nanospheres. The gratings are the interaction of the nano-object with the evanescent field. The application of the microscope that is depicted in Fig. 1.8 is the examination of human saliva in a saline solution contained in a depression well standard slide.

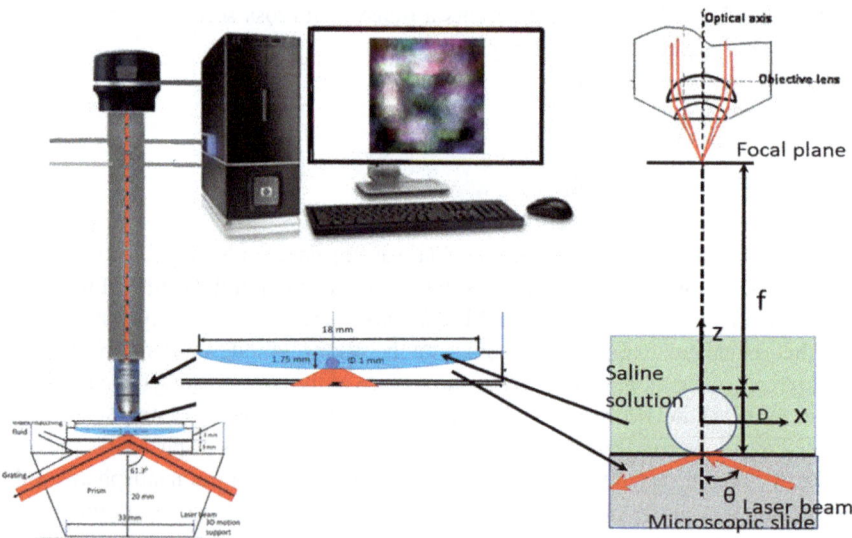

Fig. 1.8 Setup for the detection of dimensions of nano-objects utilizing the whispering gallery equatorial waves

Following the same principle utilized in our original experimental observation of nano-objects, the relay ball lens projects the observed objects into its focal plane. The microscope focus the image on the camera sensor. The digital microscope input goes to a desktop computer that contains all the required software for acquisition of images and data processing. The observed images are displayed in the desktop computer screen.

1.3.2 Role of the Carrier Gratings in the Formation of Images

The spatial resolution of an image depends on the optical resolution of the image-forming lens system and on the resolution capabilities of the utilized sensor. The optical resolution of diffraction-limited optical system is given by the Abbe constrain. It depends on the wavelength of the illuminating light and on the numerical aperture of the lens system. For practical purposes depending on the wavelength of the illuminating beam that is assumed, the limit resolution is either 220 nm or 250 nm. Applying the Nyquist condition for the spatial resolution of 220 nm means that the sampling distance of the sensor must be 110 nm or less to capture the features produced by the lens system. If one has more pixels than is required by the Nyquist criterion, the image will be oversampled, and no additional information on the observed object is gained, but the accuracy can be improved by getting redundant information.

The proposed super-resolution method increases the optical spatial resolution by making available spatial image formation frequencies not attainable with classical illumination procedures. The utilized spatial frequencies to form images can be thought in terms of a classical method utilized in the moiré method, the introduction of carrier fringes to increase the sampling frequency of an image [17]. In this application the sampling tool are carrier gratings introduced in the optical system, Fig. 1.9. It is also essential to define the concept of spatial resolution utilized in this work. Since we are measuring geometrical parameters, the concept of spatial resolution is the measurable distance between points of interest, a concept that is applied in the transmission electronic microscopy [18].

Figure 1.10a shows the image of a 6 μm polyethylene sphere that was part of the optical system that produced the observed images. Figure 1.10a shows the region of the image where the measurement was performed. Figure 1.10b shows a higher magnification image that is obtained by filtering orders in the FFT of the region under observation. In the FFT of Fig. 1.10c are present 120 orders.

The carrier gratings observed in Fig. 1.10b are harmonics of the grating present in the optical circuit, the fundamental carrier of the system.

To connect the quantities of interest in the performed measurements utilizing super-resolution, one can utilize the equation of the light intensity of the field. Calling the u(x,y) the projection of the quantity of interest in the x-axis, and for $y = y_c$, a constant, the intensity along the x-axis is,

$$I(x, y) = I_0 \left\{ 1 + \cos \frac{2\pi u(x, y_c)}{p} \right\}$$ (1.8)

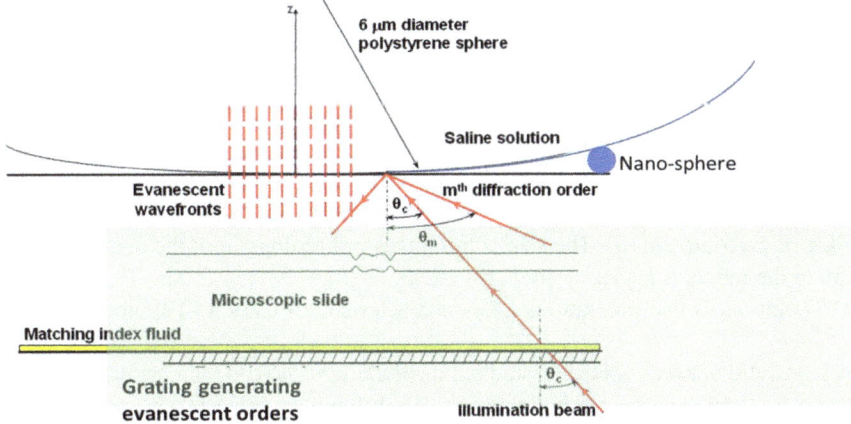

Fig. 1.9 Interface between microscopic slide and nanosphere. Detail of nanosphere position with respect to polystyrene sphere ball lens

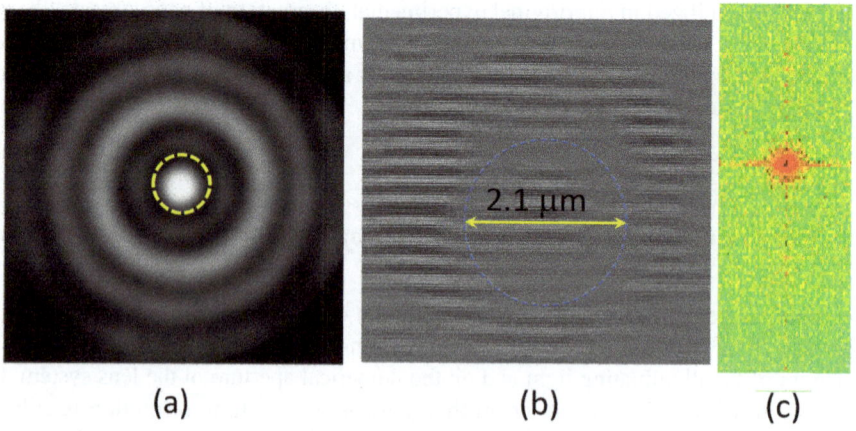

Fig. 1.10 (a) Image of the 6 μm spherical ball lens, yellow circle region where the observed nano-objects are located. (b) A view of the same region where some of the carrier gratings are shown. (c) FFT of the region shown in (a)

where p is the pitch of the selected frequency in the FFT shown in Fig. 1.10c. The value of p changes with the selected diffraction order N, then taking into consideration Eq. (1.8), we obtain a more general expression,

$$I(x,y) = I_0 \left\{ 1 + \cos \frac{2\pi u(x,y_c)}{\frac{p_o}{N}} \right\} \tag{1.9}$$

In Eq. (1.9) p_o is the pitch of the grating introduced in the optical circuit. The camera sensor is an essential element in this process since the spatial information must be captured by the sensor and transformed into a digital signal form for further processing. There are two important quantities that define the spatial resolution that can be reached based on a low-cost digital microscope that could be the basis of an inexpensive system to detect the presence of the COVID-19; an example will clarify the process. Commercially available inexpensive digital microscopes can be utilized for this purpose. These microscopes have sensors in the range of the smallest sensors sizes that can be currently manufactured, for example, 1600×1200 square pixels of size 1.12 μm. According with these specifications a line-pair is.

$2 \times 1.12 = 2.24$ μm. Utilizing the above sensor and with a magnification of 200 and using a grating of pitch $p_o = 50$ nm, the line-pair dimension is 11 nm. The pixel value is then $11/2 = 5.5$ nm. Selecting the fifth harmonic of the grating, $N = 5$ it is possible to get a pixel size of 1.1 nm. The selection of the pixel size depends on the smallest detail that we want to retrieve in the image. We have shown that it is possible to provide a solution for a particularly important problem that humanity is currently facing, an accurate, inexpensive, and fast process to detect the presence of the COVID-19 virus in a person in real time.

1.3.3 Structure of the COVID-19 Virus

In this section we will describe the geometry of the COVID-19 virus that will enable us to obtain the necessary information via this optical approach [19].

As shown in Fig. 1.11, the COVID-19 19 virus is roughly a spherical shell, and it has a membrane with an average diameter of 85 nm, and an average thickness of 7.8 nm. As such, the ratio of the thickness to radius sphere $t_{th}/r_{sp} = 7.8/42.5 = 0.1835$ is that of a thin shell; inside the membrane is the genetic material of the virus. Attached to the membrane there are several proteins and major spikes of glycoprotein (S). The size of the spikes protruding out of the membrane is 14 nm; the ratio of the length of the protrusion to the radius is $l_{spi}/r_{sp} = 14/42.5 = 0.33$.

What is important to point out is that this ratio is a key characteristic of COVID-19; other corona viruses have different ratios.

In Fig. 1.12b the average ratio between spike size and radius of the virus is 0.34, against the previously mentioned value of 0.33 obtained from a more comprehensive methodology, a 3% difference, value that is within the range of experimental verifications. The transmission electron microscope image clearly shows the virus structure. It is possible to see that sizes and shapes vary in this picture. The virus size is highly variable with average diameter of 85 nm; it can be a minimum of 50 nm and

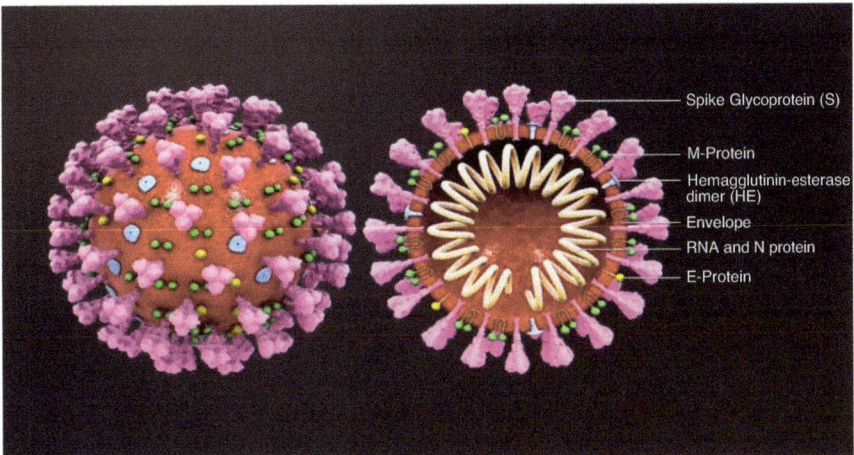

Fig. 1.11 COVID-19 structure

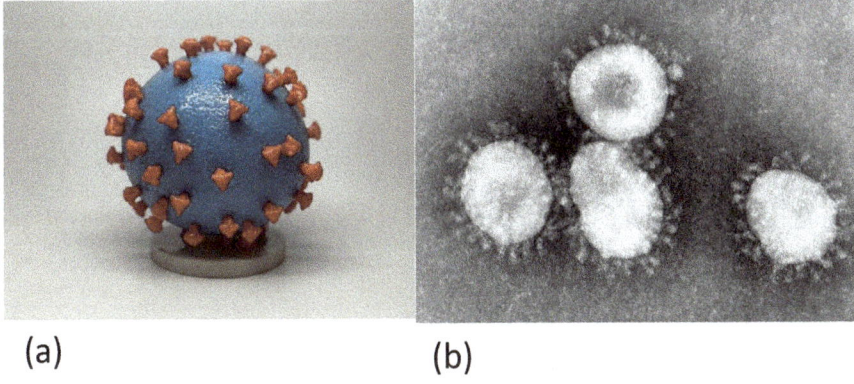

Fig. 1.12 (**a**) 3D reconstruction of the COVID-19 virus showing the spikes; (**b**) Transmission electron microscope image of COVID-19

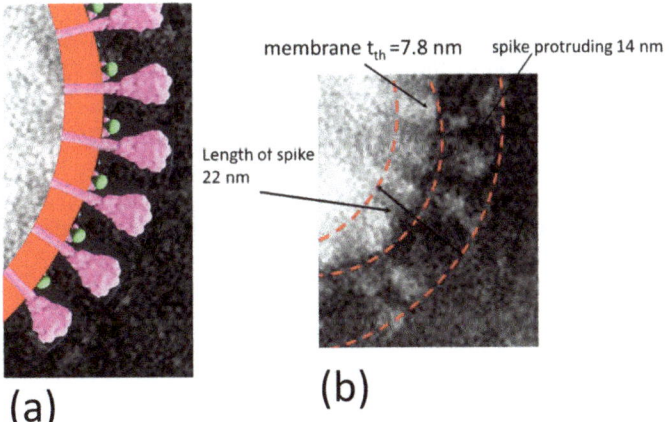

Fig. 1.13 (**a**) An enlarged image taken from Fig. 1.12b showing the membrane and the spikes approximately in scale. (**b**) Actual electron microscope image

can reach 120 nm. The envelope membrane in the transmission electron images appears as a dark region. This characteristic indicates that the membrane produces a minimum intensity in the electron microscope image.

Figure 1.13a shows the reconstruction of a region of the periphery of the COVID-19 virus. Figure 1.13b shows the corresponding image of the region as seen in the electron microscope; it is possible to see that the spatial resolution of the

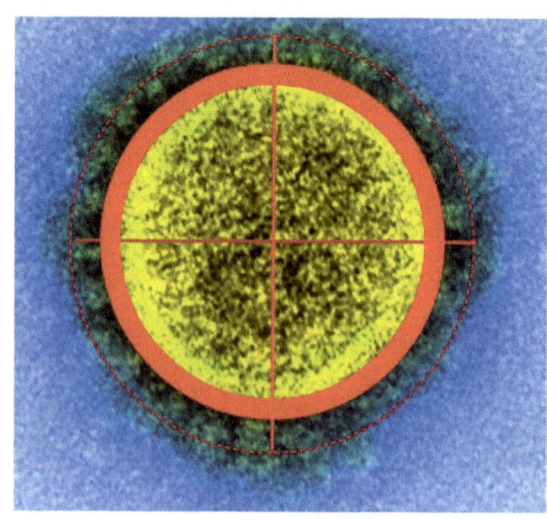

Diameter=85 nm
t_{th}=7.8 nm

Protruding spike length=14 nm
Total diameter=113 nm

Field of view 258x258 nm

Fig. 1.14 Electron microscopic image of covid-19. Main dimensions; t_{th} is the thickness of the shell

Fig. 1.15 A spike on the
COVID-19 virus

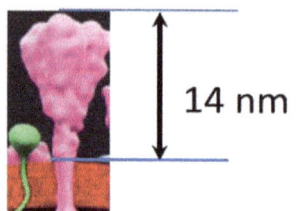

electron microscope is not good enough to define the shape of the spikes shown in Fig. 1.13a. The structure of the COVID-19 is a thin shell with prominent spikes. Figure 1.14 summarizes the main dimensions.

Figure 1.14 shows the geometry of the COVID-19 virus and the principal dimensions. An alternative mode to view the virus is to use the whispering gallery mode (WGM) of the spherical resonators described in Sect. 1.3. This methodology applies also to hollow whispering gallery resonators, optical microcavity resonators where light can travel in modes that are supported within the thin wall of the hollow structure. From observations made with dynamic electron microscopy it is possible to conclude that the connection of the spikes to the shell is equivalent to a hinge, since the spikes are in continuous oscillation motion independent of the position of the membrane. In a model of the structure of the COVID-19 virus, the material inside the protective shell can be assimilated to a fluid.

1.4 Model to Develop Techniques of Detection of the Covid-19 Virus

Figure 1.14 provides a starting point to develop a process for the detection of the virus. The cover membrane can be represented as a thin shell, a hollow spherical resonator. The shell interior can be assumed to be fluid. The spherical resonator is in a liquid medium, a saline solution that is added to saliva samples, Fig. 1.9. There are additional components, the spikes shown in Fig. 1.14, that are connected to the shell. The spikes proteins are different from the membrane proteins, as is shown in Fig. 1.15.The emission of light that is the basis of the super-resolution is connected to the vibration of the atoms producing an acoustic wave that results in the conversion of mechanical energy into electro-magnetic energy of the emitted light.

Figure 1.16a shows the standing wave of the WGM along a nanosphere diameter; Fig. 1.16b shows the image in levels of gray of the standing equatorial wave of a nanosphere of 150 nm diameter; Fig. 1.16c shows average intensity and diameter of the nanosphere. Figure 1.16d shows a color image of the same sphere.

The nanosphere is made of polystyrene that has a resonance peak at the wavelength $\lambda = 386$ nm. The wavelength of the violet color emitted light by the nanosphere is $\lambda = 386$ nm, which corresponds to UV radiation. Assuming that simulation shown in Fig. 1.18 produces patterns similar to the patterns of Fig. 1.16, a possible process of detection of the virus based on geometric measurements can be developed.

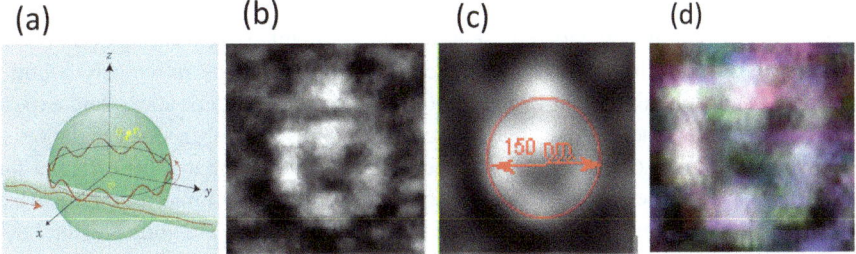

Fig. 1.16 (**a**) Whispering gallery form of oscillation of a nanosphere showing the equatorial stationary wave; (**b**) recorded pattern in levels of gray; (**c**) average intensity and diameter of the nanosphere; (**d**) color image of the same sphere

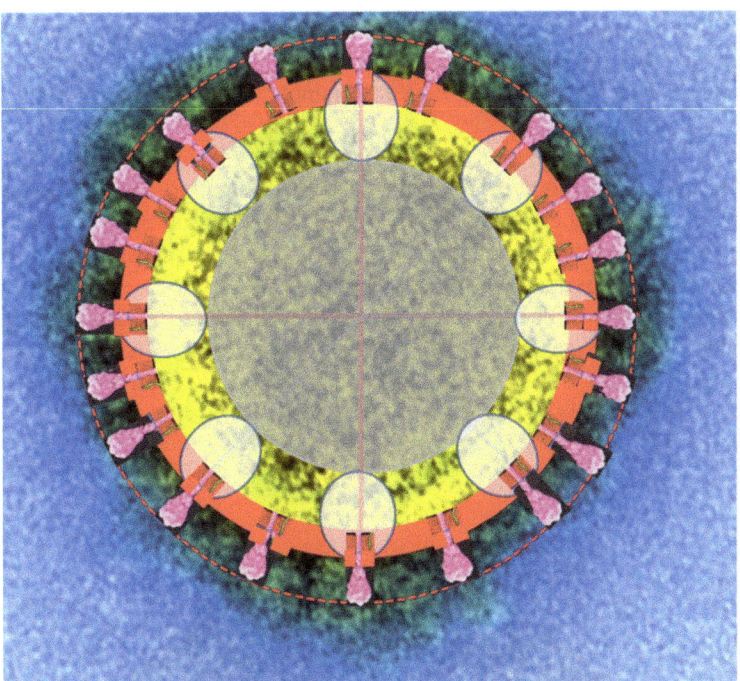

Fig. 1.17 Model of COVID-19 simulating the whispering gallery generated emission for $\lambda = 222$ nm, one of the possible emissions of the protective membrane, red area in the picture

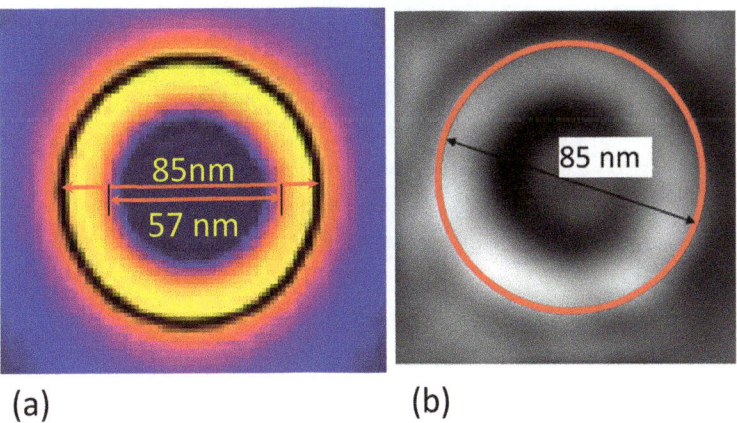

Fig. 1.18 Assumed WGM averaged intensities of nanospheres for a diameter of 85 nm. (**a**) Averages obtained from a numerical mode; (**b**) average from an actual pattern

Figure 1.17 shows a simulation of the of the WGM intensities corresponding to the COVID-19 virus, assuming the average diameter of 85 nm. The bright ellipsoidal elements represent the maximum intensities corresponding to the equatorial wave of the WGM; the maximum axis of the ellipsoid has been selected according to the previously acquired information in the whispering galley patterns of nanospheres shown in Fig. 1.16. From theoretical derivations and experimental evidence, the WGM mode of vibration of a hollow shell is like that of a corresponding sphere. In the simulation shown in Fig. 1.17 in addition of the shell, we have the spikes of the virus. We need to be considered what is the effect of the spikes in the WGM mode of vibration of the nano-shell. Experimental evidence from temporal electron microscopy tells us that the spikes are in continuous oscillation motions that are independent from the shell position. This fact indicates that the connection between shell and spikes can be assimilated to a hinge that allows the oscillation motion of the spikes. Then is highly likely that the WGM mode of vibration is not substantially affected by the spikes. Figure 1.18 shows the average light intensities corresponding to the model of WGM assumed in Fig. 1.17 and accepting that the perturbation caused by the spikes is negligible. Human saliva contains many microscopic components, possibly also including other corona viruses. Consequently, a detection method must have a reliable way to identify the COVID-19 virus and separate it from other corona viruses that may be present in the saliva samples. To detect the COVID-19 one can use dimensional parameters that characterize the COVID-19 specifically and that are different from the dimensional parameters of other corona viruses present in the saliva. This methodology can be applied to high-resolution images similar to the images shown in Fig. 1.16. Different magnifications could be required in the studies connected to COVID-19, if for example it may be necessary to make quantitative evaluations of the viral charge in a saliva sample.

There is an alternative way to detect the presence of the virus that may be applied at different magnification scales. The wavelength of the light emitted by the virus is dependent on the molecular structure of the virus shell. Lasers of diverse optical frequencies may be required to identify the type of proteins of COVID-19 protecting shell that can excite the whispering gallery mode of vibration. Consequently, this fact can produce a signature of the COVID-19 virus creating a reliable identification separating COVID-19 from other corona viruses (Fig. 1.19).

In the periphery of the virus as shown for example in Fig. 1.12 we have the spikes. It is highly likely that due to their motion, the spikes will produce fuzzy images as shown in the electron microscope images. Hence, we will have an additional ring in the image of the virus. This fact is illustrated in Fig. 1.10 where to the pattern of the shell is added a yellow ring corresponding to the presence of the spikes.

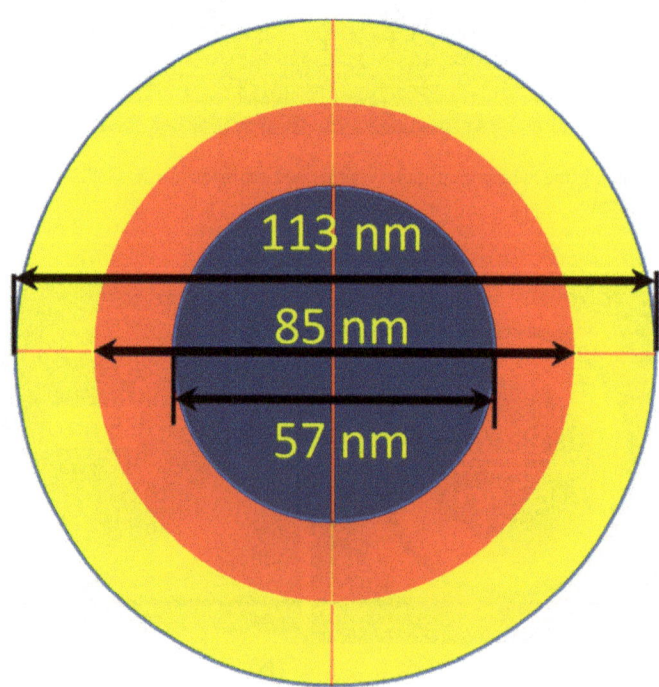

Fig. 1.19 To the parameters of Fig. 1.18, another parameter is added, the total diameter adding to the diameter of the shell the protruding length of the spikes, yellow ring

1.5 Summary and Conclusions

In the literature of developments associated with the COVID-19 pandemic, imaging technologies have a particularly important role in the process of understanding the virus behavior and the developments associated with the mitigation of the virus effects in humans. There is an extraordinary array of available tools for researchers of different aspects of the pandemic. The proposed method can have a dual role, a tool for researchers with the advantage that can be applied in environmental conditions that are close to those that ordinarily surround the virus. At the same time, the proposed method can reach extremely high spatial resolutions without interfering with the virus structure as it occurs with the electron microscope. In addition, the proposed method can be used as a diagnosis tool. Most importantly this methodology can be used to assess the severity of the disease by measuring the viral charge present in the saliva. The goal in this case is to develop as it has previously anticipated, a low-cost setup with extremely high accuracy and introducing the following software features:

1. Automatic focusing.
2. Control programs for the operation of the microscope.
3. Artificial intelligence programs to recognize requested features, and to evaluate number of these features in the analyzed volume.

The outputs should be automatic and independent of the operators' decisions.

References

1. Sciammarella, C.A., Lamberti, L., Sciammarella, F.M.: Experiment Mech. **49**, 747–773
2. Sciammarella, C.A., Lamberti, L., Sciammarella, F.M.: Digital holography to recover 3-D particle information, SEM2005. In: Proceedings of conference on Experimental Mechanics, Portland, USA (2005)
3. Sciammarella, C.A., Lamberti, L.. Optical detection of information at the sub-wavelength level. In: Proceedings of the NANOMEC06 symposium on materials science and materials mechanics at the nanoscale. Modeling, Experimental Mechanics & Applications, November 2006, Bari, Italy
4. Sciammarella, C.A., Lamberti, L.: Observation of fundamental variables of optical techniques in the nanometric range. In: Gdoutos, E.E. (ed.) Experimental Analysis of Nano and Engineering Materials and Structures. Springer, Dordrecht (2007)
5. Sciammarella, C.A., Lamberti, L., Sciammarella, F.M.: Light generation at the nano scale, key to interferometry at the nanoscale. Experiment Appl Mech. **6**, 103–115 (2010)
6. Sciammarella, C.A.: Experimental mechanics at the nanometric level. Strain. **44**(1), 3–19 (2008)
7. Sciammarella, C.A., Lamberti, L., Sciammarella, F.M.: Optical holography reconstruction of nano-objects. In: Rosen, J. (ed.) Holography, Research and Technologies, pp. 191–216. InTech, Rijeka (2011)
8. Brillouin, L.: Les électrons dans les métaux et le classement des ondes de de Broglie correspondantes. C. R. Hebd. Seances Acad. Sci. **191**, 292–294 (1930)
9. Brillouin, L.: Course De Physique Théoretique. Mason et Cie, Paris (1938)
10. Burch, J.W., Gates, C., Hall, R.G.N., Tanner, L.H.: Holography with a scatter-plate as a beam splitter and a pulsed ruby laser as light source. Nature. **212**, 1347–1348 (1966)
11. Spencer, R C., Anthony, S.A.: Real time holographic moiré patterns for flow visualization. Appl. Opt. **7**, 561 (1968)
12. Pack, A.: Current topics in nano-optics. PhD Dissertation. Chemnitz Technical University, Chemnitz (2001)
13. Bouchal, Z.: Non diffracting optical beams: physical properties, experiments, and applications. *Czechoslovak J Phys*. **53**, 537–578 (2003)
14. Hernandez-Aranda, M.: Guizar-Sicairos and M.A. Bandres. "Propagation of generalized vector Helmholtz-Gauss beams through paraxial optical systems". Opt. Express. **14**, 8974–8988 (2006)
15. Durnin, J., Miceley, J.J., Eberli, J.H.: Diffraction free beams. Phys. Rev. Lett. **58**, 1499–1501 (1987)
16. Gutiérrez-Vega, J.C., Iturbe-Castillo, M.D., Ramirez, G.A., Tepichin, E., Rodriguez-Dagnino, R.M., Chávez-Cerda, S., New, G.H.C.: Experimental demonstration of optical Mathieu beams. Opt. Commun. **195**, 35–40 (2001)
17. Sciammarella, C.A., Sciammarella, F.M.: Experimental Mechanics of Solids. Wiley, Chichester (2012)
18. Williams, D.R., Carter, B.: Transmission Electron Microscopy, a Textbook for Materials Sciences, 2nd edn. Springer, New York (2009)
19. Images of the covid-19 shown in this paper are taken from the Image Library of CCD Newsroom where the original references are provided

Chapter 2
Mesostructure Evolution During Powder Compression: Micro-CT Experiments and Particle-Based Simulations

Marcia A. Cooper, Joel T. Clemmer, Stewart A. Silling, Daniel C. Bufford, and Dan S. Bolintineanu

Abstract Powders under compression form mesostructures of particle agglomerations in response to both inter- and intra-particle forces. The ability to computationally predict the resulting mesostructures with reasonable accuracy requires models that capture the distributions associated with particle size and shape, contact forces, and mechanical response during deformation and fracture. The following report presents experimental data obtained for the purpose of validating emerging mesostructures simulated by discrete element method and peridynamic approaches. A custom compression apparatus, suitable for integration with our micro-computed tomography (micro-CT) system, was used to collect 3-D scans of a bulk powder at discrete steps of increasing compression. Details of the apparatus and the microcrystalline cellulose particles, with a nearly spherical shape and mean particle size, are presented. Comparative simulations were performed with an initial arrangement of particles and particle shapes directly extracted from the validation experiment. The experimental volumetric reconstruction was segmented to extract the relative positions and shapes of individual particles in the ensemble, including internal voids in the case of the microcrystalline cellulose particles. These computationally determined particles were then compressed within the computational domain and the evolving mesostructures compared directly to those in the validation experiment. The ability of the computational models to simulate the experimental mesostructures and particle behavior at increasing compression is discussed.

Keywords Granular material · Compression · Fracture · Discrete element method · Volumetric imaging · Image segmentation

2.1 Introduction

Recently, we have pursued development of a multi-scale, computational-experimental approach to create novel capabilities enabling process-structure-property-performance design and optimization of powder compacts. While motivated by our use of granular materials at Sandia (many of which are explosive, pyrotechnic, or energy storage materials), the challenge of predicting bulk behavior of compressed granular materials is applicable on a much larger scale. Successful prediction of the compaction behavior of a granular material requires sub-models for individual particle strength and fracture, and representations of the distribution of particulate size, morphology, and surface characteristics. Here, we build upon our prior research using an exemplar granular material of microcrystalline cellulose (MCC), adding 3-D imaging by micro-computed tomography (micro-CT) to our data of uniaxial, confined compression [1] and confined compression with 2-D imaging [2]. These experimental efforts use discrete element method (DEM) [3] and peridynamics [4] to aid in the development of computational workflows, creating new capabilities to model granular materials in compression.

While others have utilized X-ray methods to interrogate local behavior in compressed granular materials [5, 6], we do so to generate image sets suitable for assessment and validation of our DEM and peridynamic approaches. We specifically seek to simulate particulate behavior through the processes of particle rearrangement, local elastic and plastic deformation, and fragmentation, which are all present to varying levels as determined by particle characteristics of strength, size, and shape.

M. A. Cooper (✉) · D. C. Bufford
Explosive Technologies, Sandia National Laboratories, Albuquerque, NM, USA
e-mail: macoope@sandia.gov

J. T. Clemmer · D. S. Bolintineanu
Engineering Sciences, Sandia National Laboratories, Albuquerque, NM, USA

S. A. Silling
Computing Research, Sandia National Laboratories, Albuquerque, NM, USA

© The Society for Experimental Mechanics, Inc. 2022
S. L. B. Kramer et al. (eds.), *Thermomechanics & Infrared Imaging, Inverse Problem Methodologies, Mechanics of Additive & Advanced Manufactured Materials, and Advancements in Optical Methods & Digital Image Correlation, Volume 4*, Conference Proceedings of the Society for Experimental Mechanics Series, https://doi.org/10.1007/978-3-030-86745-4_2

To model the compaction of MCC particles, we used two separate techniques: DEM and peridynamics. In both methods, solids are represented using a meshfree formulation that naturally allows for cracks and discontinuities in the system. Solids are coarse-grained and represented by a collection of nodal points that exchange forces through two or many body interactions. Peridynamics is the higher fidelity method, as it is designed to solve the continuum mechanical equations of stress and strain, while DEM is the lower fidelity, but faster option, which effectively connects nodal points with breakable springs to produce an elastic response and fracture at high stresses. We recreate the experimental geometry using micro-CT imaging in both methods and simulate compaction. Results are compared to experiments with the goal of qualitatively reproducing compaction curves and macroscopic behavior, such that models can be calibrated and used to explore a larger set of problems.

2.2 Experiment

A custom die compression apparatus was developed for use within a Zeiss Xradia 520 Versa micro-CT instrument. A commercial mechanical loading frame (Debon 5KN) is available for the Zeiss Xradia instruments; however, our specific instrument configuration does not leave space for the commercial system, and our future plans include testing with hazardous materials like explosives, which led us to develop a compact custom solution. Images and schematics of our custom apparatus appear in Fig. 2.1. While the design shares features with both commercial devices and our benchtop compression apparatus [1, 2], the major differences that were necessary to optimize imaging volume within our micro-CT instrument include the lack of simultaneous force measurements and the stepwise displacement of the upper ram.

Central to the compression apparatus is a cylindrical die body made from a high-strength machinable ceramic (Macor®, Corning, Inc.) with density low enough to allow imaging by X-ray transmission. Imaging of the sample material occurs in the center section, which has an outer diameter of 3.18 cm (1.25 in), a height of 4.45 cm (1.75 in), and an inner diameter of 0.643 cm (0.253 in) allowing for a slip fit with the compression rams. The ends of the Macor die body are 1.27-cm (0.5-in) thick with a 4.32-cm (1.7-in) outer diameter and are clamped to a base plate at the bottom and a loading plate at the top. The upper and lower rams are Class ZZ gage pins (type 0.253-inch). Vespel® (DuPont) pads were added to the tip of the compression rams to alleviate beam hardening artifacts that occur at interfaces between materials of different densities or between dense materials and air.

The base plate includes a machined flat that is used to precisely align the apparatus each time it is reinserted into the micro-CT machine. Centered on, and fastened to, the base plate is a stainless steel 304 lower clamp that is used to fixture the bottom end of the Macor die body and support the bottom ram. A similar clamping scheme is used at the top of the Macor die body fixturing it to the loading plate that contains the drive bolt. The drive bolt is a 1/2-20 grade 8 bolt with a blind hole machined into the threaded end for mating with the upper ram. A 0.635-cm (0.250-in) diameter tungsten ball was positioned at the interface between the drive bolt and upper ram to prevent the upper ram from spinning within the die body as the drive bolt was turned. Due to friction at the interfaces between the upper ram, the Vespel tip, and the particulate sample at high strains, rotation of the upper ram was not always prevented.

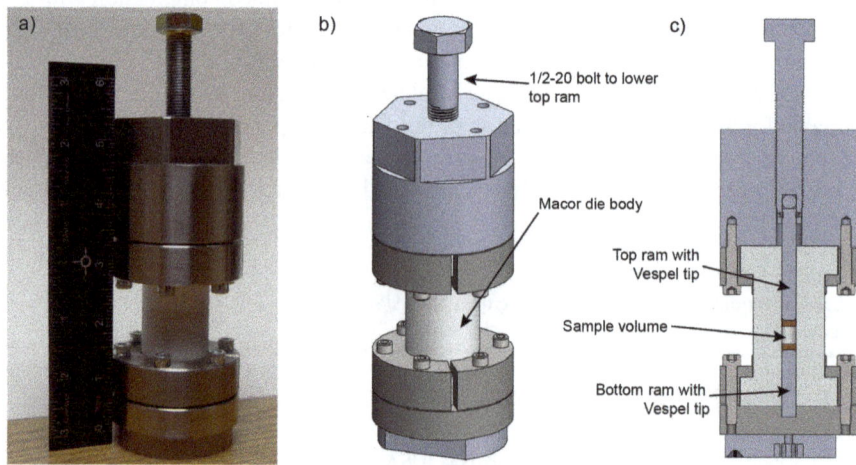

Fig. 2.1 Overview image of the compression apparatus. **a)** Photograph, **b)** solid-model illustration, and **c)** section view showing the internal compression rams

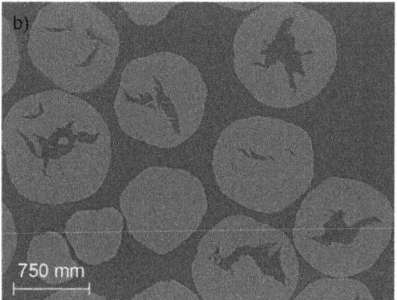

Fig. 2.2 Vivapur MCC Spheres of type 1000. (**a**) Scanning electron microscope image showing nearly spherical shape and (**b**) micro-CT volume slice showing particles with a single, large internal void

When the top and bottom ram are in contact (e.g., no sample installed), the total apparatus height is 18.3 cm (7.2 in). The drive bolt pitch results in advancing the ram by 0.213 ± 0.051 mm (0.0084 ± 0.0020 in) with 60 degrees of rotation in a displacement-controlled fashion. Under displacement control, the material remains stable outside of stress relaxation processes, which minimizes any potential movement during the hours-long micro-CT scan times. By contrast, a load-controlled approach might allow for creep during the scan interval. Eleven scans were conducted with the X-ray source operated at 160 kV and a pixel size of 3.60 μm. Each scan required approximately 24 hours to complete. The ram was advanced after each successive scan. The Macor die body was initially white, but developed a yellow/brown color with increasing X-ray exposure. Aluminum and magnesium oxides are known to accumulate optically active defects (color centers) upon X-ray exposure; Macor includes both compounds, so it is likely that such defects caused the observed color change.

Sample particles consisted of Vivapur® MCC Spheres of type 1000 from JRS PHARMA (Weissenborn, Germany. Batch No. 5100070317X). Particle size distributions were measured with a Beckman Coulter LS 13320 laser diffraction particle size analyzer with 50% of the particle size distribution equal to 1.163 ± 0.128 mm. X-ray diffraction analysis primarily found the cellulose Iβ polymorph, with average crystallite size of around 5 nm. The manufacturer reports crystallinity in the range of 70-90%. The particle shapes are roughly spherical and have internal porosity typically in the form of a single void as shown in the images of Fig. 2.2. Water content in the particles was measured by drying under vacuum at 105 °C for 24 hours and was nominally constant at 4.5%. Literature reports an MCC elastic modulus value of 7.5 GPa [7]. Our previous work [1] reported a range of fracture strengths determined by conducting unconfined single particle compressions. These strengths and the associated load-displacement curves were used to provide realistic materials responses for the DEM and peridynamics models.

In the absence of load cells, the applied stresses are inferred from benchtop compression data, matching equivalent strain. We note that these inferences are facilitated by the similar dimensions of the micro-CT apparatus and our benchtop compression apparatus. In our benchtop compression apparatus, displacement of the upper ram is pneumatically controlled, and two load cells measure forces across the powder bed [1]. The diagnostic measurements can be converted into continuous records of relative density or strain as a function of applied stress (Fig. 2.3). New data was collected with a Macor die body in the benchtop compression apparatus to assess wall friction. When prior data with a stainless steel 304 die body [1] is compared to new data, the Macor die body shows a slightly lower relative density and strain at a given applied stress. This suggests that wall friction is larger in the micro-CT compression apparatus due to the different confinement material. The lower strength of the Macor die body, as compared to the stainless steel 304 die body, limited the maximum applied stress in the benchtop compression apparatus. The data for the Macor die body in Fig. 2.3 was used to infer the bulk stress from the bulk strain measured in the micro-CT compression apparatus.

2.3 Volumetric Reconstruction, Segmentation, and Particle Library

The reconstructed micro-CT images constitute a three-dimensional grayscale image (Fig. 2.4a), where grayscale intensity corresponds to X-ray attenuation at a given spatial location. Two levels of segmentation are relevant here: first, segmentation of pore and solid phases (sometimes referred to as semantic segmentation), which results in a binarized image where each voxel is uniquely assigned to the pore or solid phase. Second, an additional level of segmentation is needed to identify individual particles (sometimes referred to as instance segmentation), where a unique identifier is assigned to each voxel to indicate to which particle it belongs. We use a random walk algorithm [8] as implemented in the open-source scikit-image

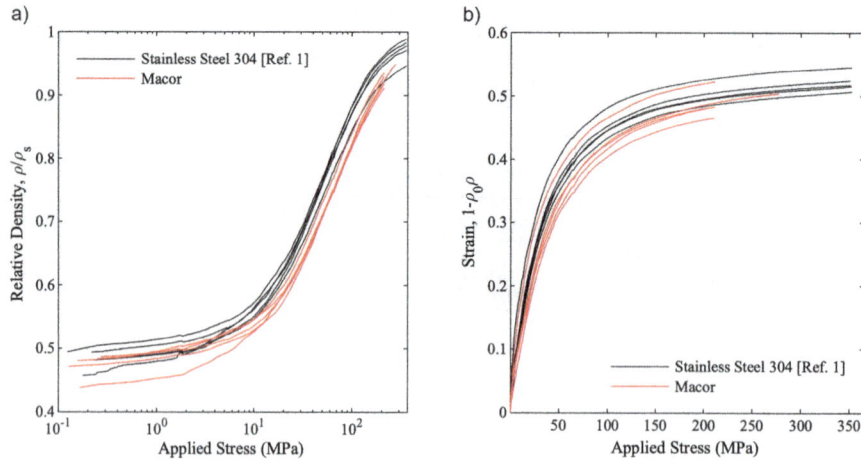

Fig. 2.3 Bulk compression data collected in the benchtop compression apparatus for a stainless steel 304 [1] and Macor die body

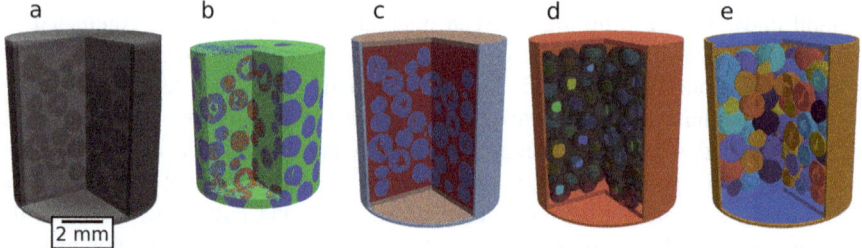

Fig. 2.4 Segmentation of reconstructed micro-CT images to extract individual particles. Three-dimensional renderings are as follows: grayscale reconstruction (**a**), label seeds for two phases of interest for random walk segmentation algorithm (**b**), random walk results for separation of solid and pore phases (**c**), label seeds for individual particles (**d**), and final segmentation of individual particles (**e**). In all cases, wedge regions are removed to allow visualization of powder bed interior

Python library [9] for both steps. First, a simple global threshold is applied to approximately separate the pore and solid phases (Fig. 2.4b), where thresholds for each phase are conservatively selected based on the intensity histogram to ensure that resulting labels are only assigned to regions that unambiguously belong to each phase (orange for solid phase, green for pore phase in Fig. 2.4b). The remaining voxels (blue in Fig. 2.4b) are then assigned via the random walk segmentation algorithm, using the partial labels as random walk seeds (see [8] for details). This results in a high-quality segmentation of pore and solid phases, Fig. 2.4c. Due to intensity variations near the walls of the apparatus, these labeling and segmentation processes are repeated iteratively for different regions near the walls, both vertically and radially (intermediate images not shown). To separate individual particles, internal porosity within each particle is first detected, using a simple connected components algorithm, and filled. A series of binary erosions are then performed on the solid phase to separate individual regions near the centers of particles, which are then assigned individual labels using a second connected components algorithm (colored regions near particle centers in Fig. 2.4d). A second random walk segmentation is performed using these individual particle labels as seeds, and the original grayscale image as an intensity map [8]. This results in assignment of individual labels to each particle in a way that respects intensity gradients (i.e., object boundaries) near particle contacts. Finally, internal porosity is re-introduced, based on the original binary segmentation, to produce the final segmented image in Fig. 2.4e.

From the final result of the segmentation process (Fig. 2.4e), individual particles can be identified, forming a virtual library of particles that can be used for multiple realizations of a simulated particle filling, particle settling, and axial compression workflow.

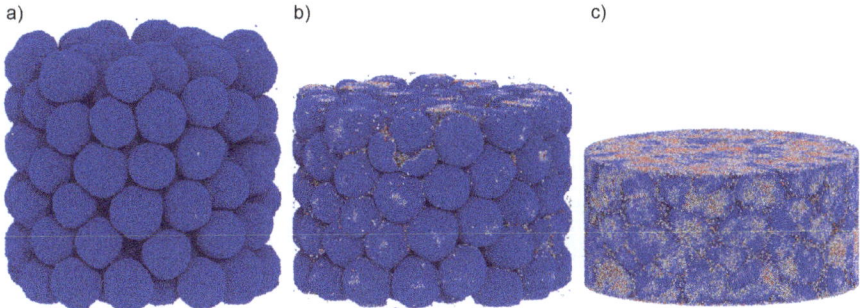

Fig. 2.5 Rendered images of compaction using DEM for relative densities of 0.47, 0.62, and 0.91 for the images from left to right. Particles are colored by number of broken bonds

2.4   Particle Compression by DEM

Discrete element modeling is a general technique used to represent granular material as a collection of small elements. These elements may be different shapes and sizes and interact through a variety of forces. Commonly, each element represents a single grain, and forces typically represent linear elastic responses or coarse-grained friction models. At high pressures, the accuracy of this approach breaks down as actual grains deform or fracture—as demonstrated in previous DEM simulations of bulk compression compared to experimental compaction in an optically accessible apparatus [2]. In this work, solid grains are represented using a bonded DEM formulation where a single grain is created by a collection of small elements or nodes [10]. Elements are point particles that interact with all neighbors using a pairwise repulsion. In addition, elements within the same grain are connected by attractive bonds that can break if stretched past a critical strain. A similar formulation is described in [11].

This model was calibrated to match the elastic and fracture properties of the MCC particles described above. Micro-CT images were used to reproduce the experimental geometry. Each MCC particle was constructed using around 20,000 nodes, as seen in Fig. 2.5. The reconstructed packing of particles was then compacted using cylindrical repulsive Lennard-Jones walls. During compaction, damage first accumulates on particle contacts, either with walls or other particles, as bonds break. As stress increases, individual particles fracture (as seen in the middle frame of Fig. 2.5) as the system densifies.

2.5   Particle Compression by Peridynamics

The peridynamic method is a generalization of the standard theory of solid mechanics that uses integral equations, rather than differential equations, permitting discontinuities such as cracks to be simulated without the need for special techniques. The method is implemented in a meshless Lagrangian discretization. The method has become widely used for applications that require simulation of many mutually interacting cracks. Peridynamics has been applied to the deformation and fracture of micrometer-scale organic crystals [12]. General information about the method can be found in [13].

For application to powder compaction, peridynamics offers the advantages of generality in material modeling, including inelastic response, as well as a simple contact algorithm. The typical response of a single MCC particle in the peridynamic model is shown in Fig. 2.6. The figure shows the compression of a particle between two plates (the top plate is not shown). The graph of load vs. plate displacement shows the features of fracture and recompression at large values of displacement.

We conducted simulations of a particle ensemble initialized from micro-CT data. The model includes a plastic yield strain of 0.01 and a tensile failure strain of 0.01. The bulk modulus is 5.0 GPa. The numerical grid is shown in Fig. 2.7 for three values of strain.

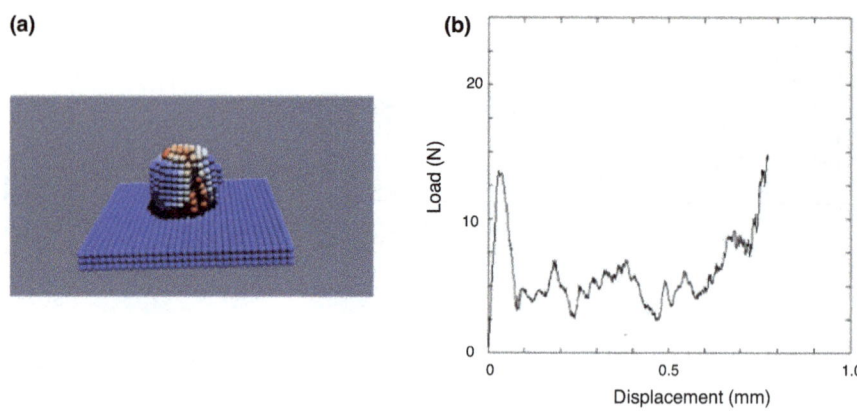

Fig. 2.6 Peridynamic model of the compression of a single MCC grain (upper plate is not shown). Left: fracture. Right: load history

Fig. 2.7 Peridynamic simulation of powder compaction initialized from micro-CT image data. Relative densities equal 0.47, 0.62, and 0.94, respectively

2.6 Discussion

While extensive statistical comparisons between mesostructure in this relatively small imaging volume were not pursued here, the demonstrated segmentation into individual particles of the experimental volume and the particle-based simulation methods suggest any number of future metrics could be extracted and compared (e.g., particle coordination number, contact areas, morphology changes). Here, we note that both simulation methods are able to represent particle phenomena of elastic and plastic deformation and fracture. Improving the quantitative comparisons for mesostructure metrics is an area of ongoing study.

Additional comparison between experiments and simulation exists in terms of the bulk compression curve of relative density versus applied stress. The experimental data of Fig. 2.3 for the Macor die body is compared to computational compression curves from the DEM and peridynamic approaches and plotted together in Fig. 2.8. In general, both computational methods show reasonable agreement with the experimental data. Potential areas for refinement include the specific material properties, such as the assumed value of 5.0 GPa bulk modulus. In fact, measurements have shown a range of values extending up to 10.0 GPa [14, 15]. The DEM utilizes an arbitrary stiffening response and relatively soft wall stiffness, both of which could be improved with additional experimental data.

The compression curve of Fig. 2.8 is plotted on a linear-log scale and change in slope is observed between a low-stress region (<10 MPa) and a high-stress region (>30 MPa). When plotted in this manner, the transition between these two nearly linear regions has been cautiously attributed to particle phenomena of fracture [16]. Notably, it is within this region that the simulated compression curves from both methods exhibit a jaggedness that results from computationally predicted particulate fracture occurring over a range of applied stresses.

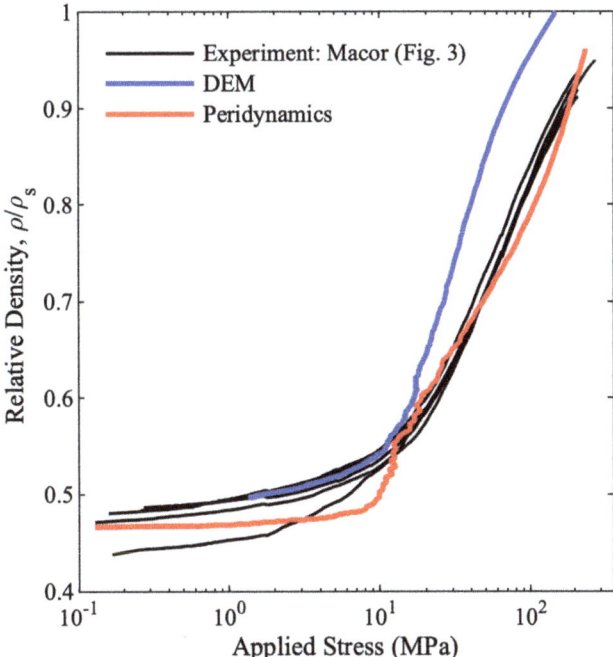

Fig. 2.8 Experiment and simulation comparison in terms of bulk compression. Black lines are the experimental data for the Macor die body from Fig. 2.3. Red line is from peridynamics. Blue line is from DEM

2.7 Summary

The major output of this study is demonstrating the workflow of experimental imaging, development of computational realizations based on real particle shapes and material strength properties, and the prediction of bulk compression behavior by simulating processes at the particle scale. To aid in model development and provide validation type data, a custom apparatus for capturing micro-CT images was developed and used to compress MCC spheres. Experimental data of qualitative particle behavior and quantitative data of bulk stress-strain during compression were compared to simulations from DEM and peridynamic methods. Both computational methods demonstrated good agreement with the experimental bulk compression stress-strain behavior.

Perhaps most exciting is the ability to perform multiple realizations of the compression process with computational domains that extract real particle shapes and sizes from a particle library. The workflow presented here was initialized with the same particle arrangement as the initial micro-CT volume; however, a straightforward extension is to perform multiple compression simulations using a randomized initial particle pouring process followed by the compression process. In this manner, the ability to optimize the compression process through multiple realizations in digital space is noted, enabling a reduced reliance on simplistic empirical relationships of bulk compression behavior.

Acknowledgments The authors gratefully acknowledge Michael Oliver for the micro-CT compression apparatus design and operation, and Andres Chavez for performing all of the micro-CT scans. This work is funded by Sandia's Laboratory Directed Research and Development program. Sandia National Laboratories is a multimission laboratory managed and operated by National Technology & Engineering Solutions of Sandia, LLC, a wholly owned subsidiary of Honeywell International Inc., for the U.S. Department of Energy's National Nuclear Security Administration under contract DE-NA0003525. This paper describes objective technical results and analysis. Any subjective views or opinions that might be expressed in the paper do not necessarily represent the views of the U.S. Department of Energy or the United States Government.

References

1. Cooper, M.A., Oliver, M.S., Bufford, D.C., White, B.C., Lechman, J.: Compression behavior of microcrystalline cellulose spheres: single particle compression and confined bulk compression across regimes. Powder Technol. **374**, 10–21 (2020)
2. Cooper, M.A., Clemmer, J.T., Oliver, M.S., Bolintineanu, D.S., and Lechman, J.B.: Visualization and simulation of particle rearrangement and deformation during powder compaction. Society of Experimental Mechanics Annual Meeting (2020)

3. Plimpton, S.: Fast parallel algorithms for short-range molecular dynamics. J. Comput. Phys. **117**, 1–19 (1995)

4. Silling, S.A., Askari, E.: A meshfree method based on the peridynamic model of solid mechanics. Comput. Struct. **83**(17-18), 1526–1535 (2005)

5. Hurley, R.C., Lind, J., Pagan, D.C., Akin, M.C., Herbold, E.B.: In situ grain fracture mechanics during uniaxial compaction of granular solids. J. Mech. Phys. Solids. **112**, 273–290 (2018)

6. Chen, Y., Ma, G., Zhou, W., Wei, D., Zhao, Q., Zou, Y., Grasselli, G.: An enhanced tool for probing the microscopic behavior of granular materials based on X-ray micro-CT and FDEM. Comput. Geotech. **132**, 103974 (2021)

7. Jonsson, H., Frenning, G.: Investigations of single microcrystalline cellulose-based granules subjected to confined triaxial compression. Powder Technol. **289**, 79–87 (2016)

8. Grady, L.: Random walks for image segmentation. IEEE Trans. Pattern Anal. Mach. Intell. **28**(11), 1768–1783 (2006)

9. van der Walt, S., Schönberger, J.L., Nunez-Iglesias, J., Boulogne, F., Warner, J.D., Yager, N., Gouillart, E., Yu, T.: scikit-image: Image processing in Python. Peer J. **2**, e453 (2014). https://doi.org/10.7717/peerj.453

10. Lisjak, A., Grasselli, G.: A review of discrete modeling techniques for fracturing processes in discontinuous rock masses. J. Rock Mech. Geotech. Eng. **6**, 301–314 (2014)

11. Clemmer, J.T.: Scale Invariant Dynamics of Interfaces and Sheared Solids. Johns Hopkins University, Baltimore (2019)

12. Silling, S., Barr, A., Cooper, M., Lechman, J., Bufford, D.C.: Inelastic peridynamic model for molecular crystal particles. Comput Particle Mech. **53**, 1047–1071 (2021)

13. Madenci, E., Oterkus, E.: Peridynamic Theory and Its Applications. Springer, New York (2014)

14. Schmalbach, K.M., Lin, A.C., Bufford, D.C., Mara, N.A.: Nanomechanical mapping and strain rate sensitivity of microcrystalline cellulose. J Mater Res. **36**, 2251 (2021)

15. Wunsch, I., Finke, J.H., John, E., Juhnke, M., Kwade, A.: A mathematical approach to consider solid compressibility in the compression of pharmaceutical powders. Pharmaceutics. **3**, 121 (2019)

16. Kenkre, V.M., Endicott, M.R., Glass, S.J., Hurd, A.J.: A theoretical model for compaction of granular materials. J. Am. Ceram. Soc. **79**(12), 3045–3054 (1996)

Chapter 3
Surface Pressure Reconstruction in Shock Tube Tests Using the Virtual Fields Method

R. Kaufmann, E. Fagerholt, and V. Aune

Abstract This study investigates full-field, dynamic pressure reconstruction during shock-structure interactions using optical measurements and the virtual fields method (VFM). Shock wave impacts pose severe challenges to experimental measurement techniques due to the substantial, almost instantaneous pressure rises they induce. Their effects are typically measured pointwise using pressure transducers or as total force using load cells. Here, surface deformations were measured on the blind side of a flat steel plate in pure bending using a deflectometry setup. Pressure was reconstructed from the deformations induced by an impacting shock wave using a piecewise VFM approach. Different shock wave symmetries were used in order to investigate the capabilities of identifying spatial distributions reliably under the experimental conditions in the shock tube. Pointwise pressure transducer measurements allowed a validation of the results. It was found that different shapes of load distributions on the sample surface can be identified qualitatively, but that the comparability of both measurement techniques is limited due to filter and sampling capabilities.

Keywords Pressure reconstruction · Virtual fields method · Full-field measurement · Shock wave · Fluid-structure interaction

3.1 Introduction

Shock tube experiments are favourable to obtain a better understanding of shock-structure interaction and may therefore be used in the development of resilient structures prone to extreme loading events. In order to improve the design of protective and resilient structures, both material behaviour and fluid-structure interactions occurring during extreme loading events are subject to current research [1, 2]. While structural responses can often be measured in full-field using DIC, load distributions during impact are usually assessed pointwise using pressure transducers. This does not allow detailed investigations of the fluid-structure interactions due to the resulting low spatial resolution of pressure data and the intrusive nature of such transducers. Available full-field techniques are generally limited in terms of their applicability, as discussed in the following. If the flow field is accessible, particle image velocimetry (PIV) and particle tracking velocimetry (PTV) allow full-field pressure reconstruction in the flow over a large range of pressure amplitudes [3, 4]. In extreme flow environments involving compressible effects like shock waves, these methods are generally not applicable because the gas bubbles that are used as tracer particles burst and only a limited number of window elements can be installed for optical access. Pressure-sensitive paints [5, 6] are often used for the measurement of large differential pressure amplitudes surfaces, but they also require optical access to the pressure side of the structure under consideration, which is often challenging. New approaches are required to achieve full-field measurements of surface loads in extreme flow environments. Addressing this gap, the present study aims at applying an approach for optically measuring deformations and then reconstructing the load distributions occurring during the impact of a shock wave on a flat, thin steel plate. Using thin plate samples in pure bending allows employing the Kirchhoff-Love theory which relates the involved deformations to the acting loads [7]. The main limitation of this method is that it

R. Kaufmann (✉)
Structural Impact Laboratory (SIMLab), Department of Structural Engineering, NTNU – Norwegian University of Science and Technology, Trondheim, Norway
e-mail: rene.kaufmann@ntnu.no

E. Fagerholt · V. Aune
Structural Impact Laboratory (SIMLab), Department of Structural Engineering, NTNU – Norwegian University of Science and Technology, Trondheim, Norway

Centre for Advanced Structural Analysis (CASA), NTNU, Trondheim, Norway

© The Society for Experimental Mechanics, Inc. 2022
S. L. B. Kramer et al. (eds.), *Thermomechanics & Infrared Imaging, Inverse Problem Methodologies, Mechanics of Additive & Advanced Manufactured Materials, and Advancements in Optical Methods & Digital Image Correlation, Volume 4*, Conference Proceedings of the Society for Experimental Mechanics Series, https://doi.org/10.1007/978-3-030-86745-4_3

involves fourth-order spatial derivatives and second-order temporal derivatives of the deflections, which tend to significantly amplify experimental noise. This issue can be addressed by using the virtual fields method (VFM). The VFM is an application of the principle of virtual work and relies on using full-field deformations and material constitutive mechanical parameters to extract load information or vice versa [8]. In case of thin plates in pure bending, employing the principle of virtual work reduces the required order of spatial derivatives of deflections to two. The VFM requires the choice of appropriate virtual fields which are subject to current research. The present study utilizes a VFM approach based on piecewise virtual fields in combination with deflectometry, a highly sensitive optical technique that allows measuring surface slopes [9]. These techniques were previously combined to reconstruct dynamic mechanical point loads [10]. Pressure reconstructions of air jets impinging on a thin glass mirror were investigated in a series of studies [11–13]. In these previous studies it was found that shapes, locations and time histories of the relatively small external loads of ca. 10–1000 Pa could be identified qualitatively well, but that the accuracy of the approach is highly dependent on signal-to-noise ratio, experimental bias and processing parameters. Simulated experiments were conducted in [11] to assess the accuracy of static reconstruction results. The present study applies this methodology to shock waves impacting steel plates to investigate its capabilities in this extreme experimental environment in terms of potential experimental bias from vibrations, temporal and spatial resolution. Two shock wave symmetries are used to demonstrate the performance of the technique to capture dynamic, spatial full-field distributions. The results are compared to pressure transducer measurements to evaluate the achieved accuracy in pressure amplitude.

3.2 Deflectometry

Deflectometry is an optical measurement technique for surface slopes [9]. As illustrated in Fig. 3.1, the basic setup consists of a specimen with specular reflective surface, a cross-hatched grid with printed pitch p_G, a camera and a light source. The camera is placed at an angle next to the printed grid such that it records the reflected grid in normal incidence. The angle should be minimized to avoid distortions in the recorded image. Here, both grid and camera are placed at a distance h_G from the specimen surface. As shown in Fig. 3.1, a camera pixel records the reflected grid point P when directed at point M on the specimen surface when no load is applied. If the specimen is loaded, it deforms and local changes of the surface slopes, d α, occur. For sufficiently small deformations rigid body movements and out-of-plane deflections can be neglected and the pixel directed at point M records the reflected grid point P′ when a load is applied. Phase detection algorithms allow an extraction of local phase information from grid images and further the phase shift that results from applying a load to the specimen. Here, phase maps were applied using a spatial phase-stepping algorithm [14]. A detection algorithm featuring a windowed discrete Fourier transform algorithm with triangular weighting was employed as it can suppress some harmonics and errors from miscalibration. A detection kernel size of two grid pitches is used in this study. The displacement, u, between points P and P′ is

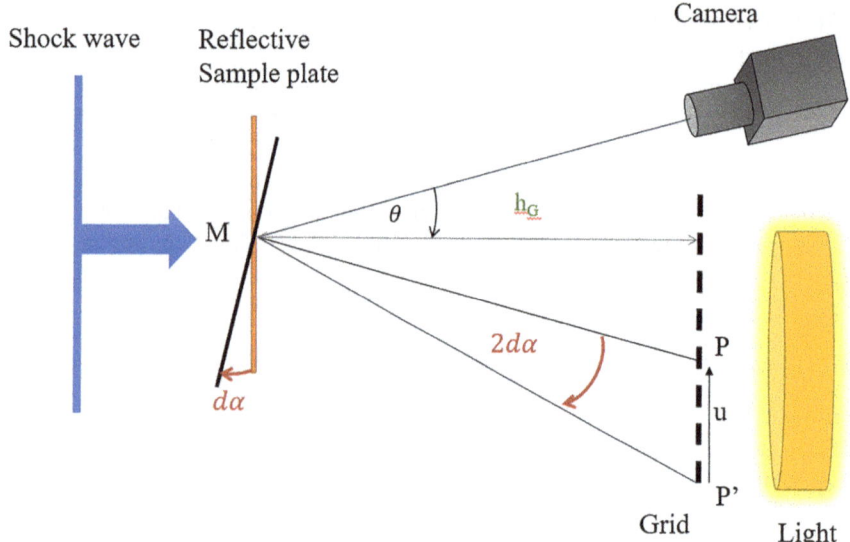

Fig. 3.1 Sketch of a deflectometry setup with specular reflective sample plate and basic principles

then obtained from the difference in phase maps between loaded and unloaded configuration. A linear relation between surface slopes d α and measured displacement u can be derived based on geometrical considerations. Assuming that h_G is large against u and the specimen dimensions, θ is negligible and that the camera records images in normal incidence, dx can be expressed as:

$$d\alpha_x = \frac{u_x}{2h_G}, d\alpha_y = \frac{u_y}{2h_G} \tag{3.1}$$

A more complicated, full calibration is required if these assumptions are not accurate [15]. The printed grid pitch, p_G, drives the resolution in both space and slope. The slope resolution is further driven by measurement noise and the distance between grid and sample surface, h_G.

3.3 Pressure Reconstruction

Based on the principle of virtual work, the VFM is a technique that allows identifying surface pressure loads from full-field kinematic measurements and known constitutive material parameters or vice versa [8]. It is computationally cheap when compared to alternative approaches like Finite Element Model updating because it does not rely on iterative procedures to match numerical and experimental results. Further, it does not require knowledge of the boundary conditions. Here, the curvatures and accelerations obtained from deflectometry measurements were used to reconstruct pressure via the principle of virtual work for a thin plate in pure bending. Eq. (3.2) describes the equilibrium of an isotropic, homogeneous plate through the principle of virtual work, assuming pure bending and linear elasticity.

$$\int_S p(t_i) \cdot w^* dS = \int_S \kappa^* D\kappa(t_i) dS + \int_S \rho t \, a(t_i) \, w^* dS \tag{3.2}$$

p is the investigated pressure, t_i a point in time, S the surface area of the element, D the bending stiffness matrix, κ the curvature and a the acceleration. w^* and κ^* are the virtual deflections and curvatures, with conditions that w^* is continuous, differentiable with continuous derivatives. Curvatures were obtained using a central difference differentiation scheme. In order to obtain accelerations, deflections were first calculated using a sparse matrix integration scheme. Acceleration were then calculated using a 5-point central difference temporal differentiation scheme. Since the material parameters are known a priori, p can then be identified once suitable virtual fields are chosen. 4-node hermite 16 element shape functions as used in the finite element method [11] are well suited here because they provide C^1 continuity of w^* and therefore C^0 continuity of the virtual slopes [9, chap. 15]. Pressure is reconstructed within a window of four hermite elements of a chosen size. Pressure is then reconstructed piecewise by shifting the window over the investigated field of view. The size of the pressure reconstruction window (PRW) is a key reconstruction parameter. Another key parameter is the slope filter kernel size. Slope filtering is conducted using a 3D gaussian filter to obtain curvature and acceleration maps that are sufficiently smooth for pressure reconstruction. The kernel size σ is defined as one standard deviation of the Gaussian. In the present study, a pressure reconstruction window with a side length of PRW = 30 data points and a slope filter kernel of $\sigma = 2$ were used. The PRW was shifted by one data point in each direction over the entire field of view. This leads to spatial oversampling of the experimental data but was found to significantly improve the results at acceptable computational cost.

3.4 Experimental Methods

The experiments were conducted in a shock tube designed for the simulation of blast waves [16]. A half blockage was added to the square nozzle exit with 300 mm side length in order to vary the symmetry of the shock wave, thus allowing an investigation of the capabilities to measure the resulting difference in pressure amplitude and history on the specimen surface. The specimen, a 300 mm square steel plate with fully clamped boundary conditions, was mounted on a rigid frame that was constructed to assure that no out-of-plane motion of the sample would occur during the experiment. Further, a wind shield was installed around the sample frame to protect the camera and grid required for the deflectometry setup. Both camera and grid were mounted on a separate frame inside the shock tube test chamber. Even though deflectometry is highly sensitive to vibrations in both camera and grid, all parasitic vibrations were assumed to occur several milliseconds after the shock wave reaches the sample plate, allowing for an accurate measurement for the duration of the initial impact on the specimen.

Table 3.1 Experimental parameters

Sample	Mirror polished steel 1.4301
Sample Young's modulus	200 GPa
Sample density	$8\ \mathrm{g\ cm^{-3}}$
Sample Poisson's ratio	0.3
Sample thickness	5 mm
Sample side length	300 mm
Sample boundary conditions	Clamped
Camera	Phantom v2511
Frame rate	75 kHz
Resolution	512 pixels × 512 pixels
Shutter speed	$5\ \mu s$
Distance camera-sample	1.37 m
Printed grid pitch	5.88 mm

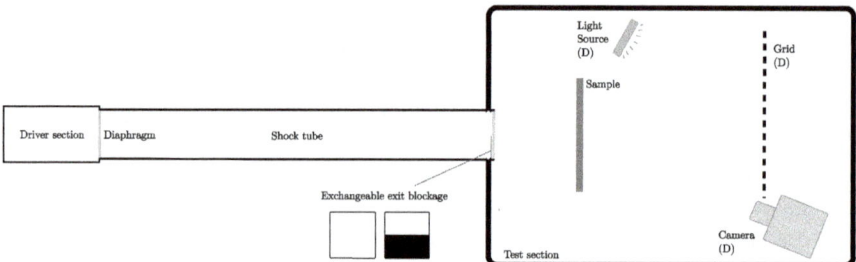

Fig. 3.2 Shock tube and deflectometry setup

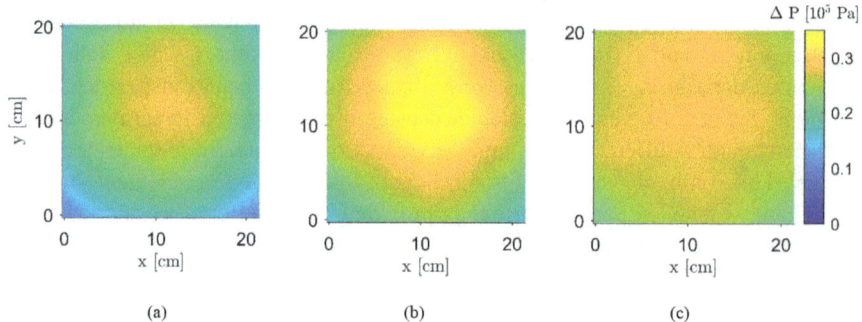

Fig. 3.3 Pressure reconstructions for open nozzle exit test case. Reconstruction at t = 0.1 ms (**a**), t = 0.14 ms (**b**), t = 0.18 ms (**c**)

Table 3.1 shows the relevant experimental parameters. Figure 3.2 shows a sketch of the setup. Additionally, pressure transducer measurements were conducted using a Kistler Type 603B sensor. For this purpose, the mirror specimen was replaced with a 2 cm thick aluminium plate with a pressure transducer placed in the centre.

3.5   Analysis

Figures 3.3 and 3.4 show reconstructed pressure maps for the two investigated cases of open and half-blocked nozzle exits. In case of the open nozzle exit, a nearly gaussian distribution is observed at all shown time steps at varying amplitude. With the bottom half of the nozzle exit blocked, the shock wave impacts the top of the plate first and induces lower peak pressure amplitudes. At later time steps, it is found that the peak pressure area propagates downwards at decreasing amplitude. The reconstructions allow an identification of the different shapes and amplitudes at all investigated time steps. It should be noted that during the time steps of initial impact shown here, inertial forces govern the dynamics, such that the accelerations extracted from deflectometry data are the most relevant quantity in terms of measurement. Figure 3.5 shows comparisons of the VFM pressure reconstructions with transducer data at the centre point. Transducer data cannot be compared directly to

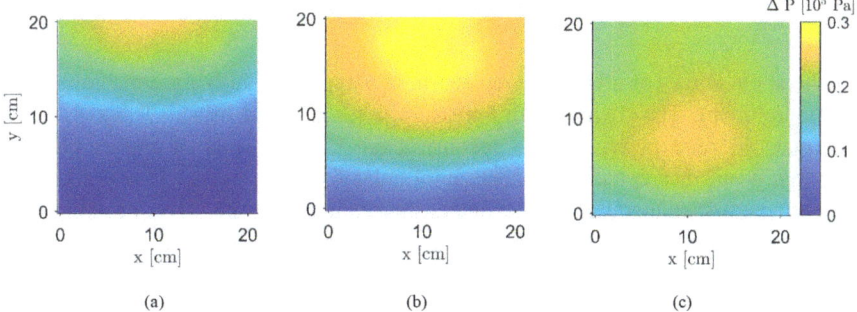

Fig. 3.4 Pressure reconstructions for half-blocked nozzle exit test case (blockage of bottom half). Reconstruction at t = 0.1 ms (**a**), t = 0.14 ms (**b**), t = 0.18 ms (**c**)

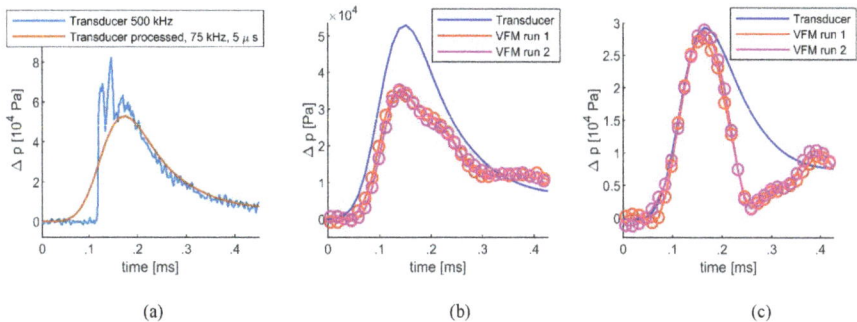

Fig. 3.5 (**a**) Original pressure transducer signal compared to processed signal to match camera frame rate of 75 kHz and shutter speed of 5 μs. Comparison between VFM and transducer data for (**a**) open and (**b**) half-blocked nozzle exit test case

optical measurement results because they record data at one point in time, while camera images are integrated over a time interval given by the shutter speed. Therefore, transducer data were recorded at 500 kHz and then averaged over intervals corresponding to the shutter speed of 5 μs. The data was then undersampled to account for the frame rate of 75 kHz. Finally, the data was filtered with a gaussian 1D filter. The latter does not fully replicate the effect of the 3D filter used on deflectometry data but should allow for a reasonable comparison. Figure 3.5a shows the original and the processed transducer signal. The originally observed, almost instantaneous rise in pressure is smoothed out, as well as some events that occur in very short time intervals.

For the open nozzle case, Fig. 3.5b shows a large discrepancy between the peak amplitudes identified with both methods. Figure 3.5c shows good agreement of peak pressures for both measurement techniques for the case of the half-blocked nozzle exit until few data point after peak pressure occurs. The amplitude reduces more quickly after the initial impact for VFM data after that point. This is likely to vibrations occurring in the setup due to the impact. The reason for the observed differences in peak amplitude for the open nozzle case is likely that the pressure transducer processing procedure makes the simplified assumption that the signal sampled at 500 kHz corresponds to a 2 μs integration time when compared to camera measurements. In practice the transducers record a single event that occurs during 2 μs almost instantaneously, thus leading to aliasing. In the half-blocked nozzle case, lower peak pressures occur such that the effect is much less noticeable. Both VFM and transducer results were found to be repeatable, such that random noise can be ruled out as reason for the discrepancy. At this point it is also important to recall that the VFM reconstruction at each point is based on a square window of 30 data points side length which are in turn being filtered with a gaussian window of 12 data points side length in both space and time. The 50 mm distance between two transducers corresponds to approximately 18 data points. The necessary space-time filtering is likely that particularly acceleration information is smeared out, and thus the governing factor for the short-time dynamics. The fact that Figs. 3.3 and 3.4 show that the pressure distributions are captured qualitatively well over several time steps for both investigated symmetries supports the argument that the deviation between transducer and VFM data is likely due to the intrinsic differences in sampling of the techniques. This issue will be investigated further in future studies.

3.6   Conclusion

This work investigates the applicability of a pressure reconstruction approach utilizing deflectometry measurements and the VFM to a shock tube environment. The shapes of the pressure fields induced by shock waves with different symmetries impacting a thin steel plate in pure bending were reconstructed. Comparisons of the amplitudes to local transducer measurements revealed significant discrepancies in peak amplitudes for some cases. This is likely because the transducer data points are recorded almost instantaneously while camera images are integrated over time intervals given by the shutter speed. Further, previous studies have found that the obtained amplitudes are very sensitive to the chosen reconstruction parameters as they influence how experimental noise and potential systematic error interfere with the signal. The 3D filter that is required to obtain sufficiently smooth acceleration maps is likely to have reduced the local acceleration amplitudes which are encoded in the short-time variations of the deflection field. This issue will be investigated further by simulating both the experiment and the measurement process. This is the only way to analyse the exact influence of filtering on the noisy acceleration and curvature maps that are used for pressure reconstruction. More transducer and deflectometry measurements will also be conducted to identify possible error sources and further explore the capabilities of the methodology presented herein.

Acknowledgements This work has been carried out with the financial support from NTNU and the Research Council of Norway (RCN) through the Centre for Advanced Structural Analysis (CASA), Centre for Research-based Innovation (RCN Project No. 237885); SINTEF Ocean and the SLADE KPN project (RCN Project No. 294748); and the Norwegian Ministry of Justice and Public Security.

References

1. Rigby, S., Tyas, A., Clarke, S.: Observations from preliminary experiments on spatial and temporal pressure measurements from near-field free air explosion. Int J Protect Struct. **6**(2), 175–190 (2015)
2. Aune, V., Valsamo, G., Casadei, F., Langseth, M., Børvik, T.: Fluid-structure interaction effects during the dynamic response of clamped thin steel plates exposed to blast loading. Int. J. Mech. Sci. **195**, 106263 (2021)
3. de Kat, R., van Oudheusden, B.: Instantaneous planar pressure determination from PIV in turbulent flow. Exp. Fluids. **52**(5), 1089–1106 (2012)
4. J. Schneiders, S. A. Caridi. F. Scarano, "Large-scale volumetric pressure from tomographic PTV with HFSB tracers," Exp Fluids 57(11): 164, 2016
5. Beverley, J., McKeon, R.: Springer Handbook of Experimental Fluid Mechanics, pp. 188–208. Springer, Berlin (2007)
6. McKeon, R.: Springer Handbook of Experimental Fluid Mechanics. Springer, Berlin, Heidelberg (2007)
7. Timoshenko, S., Woinowsky-Krieger, S.: Theory of Plates and Shells. McGraw-Hill, New York (1959)
8. Pierron, F., Gédiac, M.: The Virtual Fields Method. Extracting Constitutive Mechanical Parameters from Full-Field Deformation Measurements. Springer, New York (2012)
9. Surrel Y.: Deflectometry: a simple and efficient noninterferometric method for slope measurement (2004)
10. O'Donoughue, P., Robin, O., Berry, A.A.: Time-resolved identification of mechanical loadings on plates using the virtual fields method and deflectometry measurements. Strain. **54**, e12258 (2017)
11. Kaufmann, R., Pierron, F., Ganapathisubramani, B.: Full-field surface pressure reconstruction using the virtual fields method. Exp. Mech. **59**(8), 1203–1221 (2019)
12. Kaufmann, R., Ganapathisubramani, B., Pierron, F.: Reconstruction of surface-pressure fluctuations using deflectometry and the virtual fields method. Exp Fluids. **61**, 1–15 (2020)
13. Kaufmann, R., Ganapathisubramani, B., Pierron, F.: Surface pressure reconstruction from phase averaged deflectometry measurements using the virtual fields method. Experiment Mech. **60**, 379–392 (2019)
14. Badulescu, C., Grédiac, M., Mathias, J.D.: Investigation of the grid method for accurate in-plane strain measurement. Measurement Sci Technol. **20**, 095102 (2009)
15. Balzer, J., Werling, S.: Principles of shape from specular reflection. Measurement. **43**(10), 1305–1317 (2010)
16. Aune, V., Fagerholt, E., Langseth, M., Børvik, T.: A shock tube facility to generate blast loading on structures. Int J Protect Struct. **7**(3), 340–366 (2016)

Chapter 4
Stress Concentration Evaluation of a Plate with Symmetrical U-Notches Under Tensile Load Using TSA and a Lepton IR Camera

V. E. L. Paiva, D. G. G. Rosa, G. L. G. Gonzáles, and J. L. F. Freire

Abstract It has already been showed that Thermoelastic Stress Analysis (TSA) may be employed as an extremely useful technique in structural integrity assessment applications. Recent developments in the microbolometer technology made possible the commercial offer of very low-cost infrared cameras, one of these being the FLIR Lepton 3.5. Use of such low-cost cameras associated with commercially available software or in-house developed algorithms makes possible to localize anomalies and to determine quantitative results on the stress distribution acting on nominal and hot-spot locations in loaded structures. A further step will widely disseminate IR temperature and TSA measurements and consequent analyses into a powerful low-cost health monitoring tool. The aim of the present work is to demonstrate the use of this experimental technique in the evaluation of the stress concentration caused by a U-notch in a plate under tension load using TSA and a Lepton camera. A MATLAB in-house algorithm was developed for post-processing the measured IR signal. The achieved stress-distribution and stress-concentration results are also compared with those generated by measurement systems that integrate a commercially available software coupled to the FLIR Lepton 3.5 and to a median-cost FLIR A655sc IR camera. Moreover, the paper shows that the low-cost camera can be used in the monitoring of fatigue crack growth as well as in the determination of stress intensity factors for cracks initiated and propagated from the U-notch.

Keywords Thermoelastic stress analysis · TSA · Lepton 3.5 · Stress concentration

4.1 Introduction

Discontinuities or abrupt geometric changes in a loaded structure act as stress raisers and cause high localized stresses [1] that may fatigue damage.

Fatigue is one of the most common cause of mechanical failure in engineering components and its prevention is a major concern in structural health [2, 3]. Fatigue cracks commonly initiate at stress concentration features such as imposed designed abrupt geometric changes, initial damage caused by some manufacturing processes, or external third-order interference and can propagate during continuous operation, not necessarily causing an imminent risk, but surely impairing its functionality, accelerating its degradation, and diminishing its service life.

The use of infrared (IR) thermography, more specifically thermoelastic stress analysis (TSA) [4], as a nondestructive evaluation (NDE) technique to analyze the behavior cyclic load components is not new, being used in large scale to evaluate the delamination of composite polymer materials [5–7], crack inspection [8–12], and estimation of remaining strength of components [13–17], playing an important role in maintenance programs and in-service inspections.

In recent years, new developments in the technology of the microbolometer have enabled the emergence of very low-cost thermal camera models, such as the Lepton one. However, these low-cost models offer reduced capabilities as compared to high-end devices mainly due to the sensor spatial resolution, simpler optics, housing, and electronics. The quality of the achieved results is proportional to the investment in the sensor; nevertheless, the possibilities of application to structural health monitoring are countless [18, 19].

The present paper presents the application of the TSA technique to analyze the stress concentration generate by a U-notch in a polycarbonate plate under tension using a low-cost Lepton camera. The IR signal was post-processed using a MATLAB

V. E. L. Paiva (✉) · G. L. G. Gonzáles · J. L. F. Freire
Department of Mechanical Engineering, Pontifical Catholic University of Rio de Janeiro, PUC-Rio, Greece

D. G. G. Rosa
Department of Civil Engineering, Rua Marquês de São Vicente, Rio de Janeiro, RJ, Brazil

© The Society for Experimental Mechanics, Inc. 2022
S. L. B. Kramer et al. (eds.), *Thermomechanics & Infrared Imaging, Inverse Problem Methodologies, Mechanics of Additive & Advanced Manufactured Materials, and Advancements in Optical Methods & Digital Image Correlation, Volume 4*, Conference Proceedings of the Society for Experimental Mechanics Series, https://doi.org/10.1007/978-3-030-86745-4_4

algorithm. The results were compared with those generated by measurement systems that integrate a commercially available software coupled to the FLIR Lepton 3.5 and to a median-cost FLIR A655sc IR camera. Digital image correlation system (DIC) and finite element method (FEM) were also employed for additional evaluation and comparison. Moreover, the paper shows that the low-cost camera can be used in the monitoring of fatigue crack growth as well as in the determination of stress intensity factors for cracks initiated and propagated from the U-notch.

4.2 Material and Experimental Procedure

The material analyzed in this work was polycarbonate, which is a well-balanced material with relatively high resistance to temperature, ductility, and both mechanical and impact strengths. For this reason, it is generally considered an engineering plastic and has multiple applications.

Two types of specimens were used herein, both made of the same polycarbonate sheet. The first one was a dog bone specimen, also known as the constant radius tensile specimen, presenting a smooth stress gradient along its length due to the continuous variation of the cross-sectional area. The uniaxial tensile stress-strain curve for the material, Fig. 4.1, was determined using DIC to measure the displacement full-field while the dog bone specimen was monotonic loaded until failure. Further relevant material data are given in Table 4.1. It is relevant to notice at this point that the stress-strain curves used the minimum cross section dimensions for stress calculation and that the strain data was determined by a long virtual gage as depicted in Fig. 4.1. No attempt was made to determine equivalent strains from the triaxial strain state after the occurrence of necking, meaning that the stress-strain curves and presented mechanical property data are accurate up to the point where stresses reach their peak.

The DIC used hardware was composed of a tripod and fixtures to adapt two 2.3 megapixel CMOS cameras (Basler acA1920-155um) coupled to adjustable focal length lenses (Tamron A031 AF28-200 mm F/3.8–5.6). A calibration target (12 × 8 dots with 15 mm of grid spacing) compatible with the DIC Vic Snap and VIC-3D software (Correlated Solutions Inc.) was used for calibration.

The second type of specimen and main object of the present analysis was a bar with symmetrical and opposite U-shaped notches (thickness = 2.94 mm, width = 49.96 mm, length = 250 mm and notch radius = 9.33 mm), Fig. 4.2a. The specimens were tested in a 100kN INSTRON servo-hydraulic machine under cyclic axial load ($R = \sigma min/\sigma max = 0.1$ and $f = 1$ Hz).

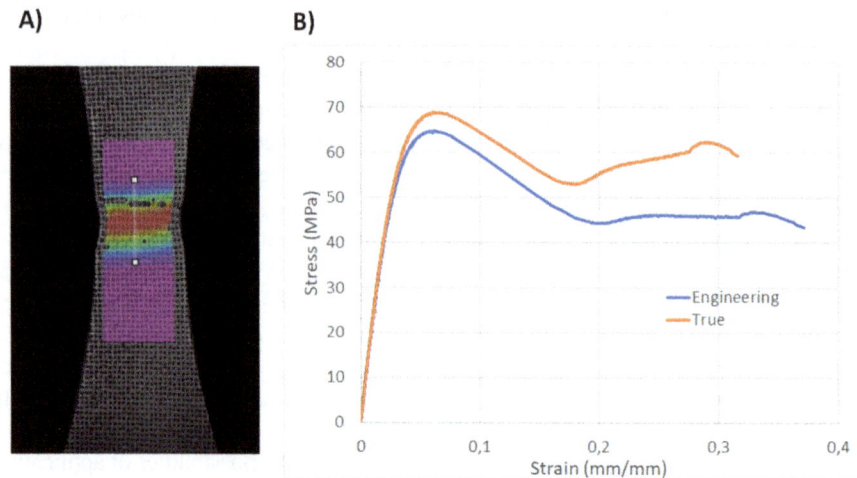

Fig. 4.1 Uniaxial tensile test: (**a**) Axial trains measured in the dog bone specimen using DIC during the test; (**b**) polycarbonate true and engineering stress-strain curves (accurate up to the stress peak)

Table 4.1 Polycarbonate engineering properties

Young's modulus	2039 MPa
Poisson's coefficient	0.42
Yield strength 0.2%	47 MPa
Ultimate (peak) strength	69 MPa
Infrared emissivity	0.9

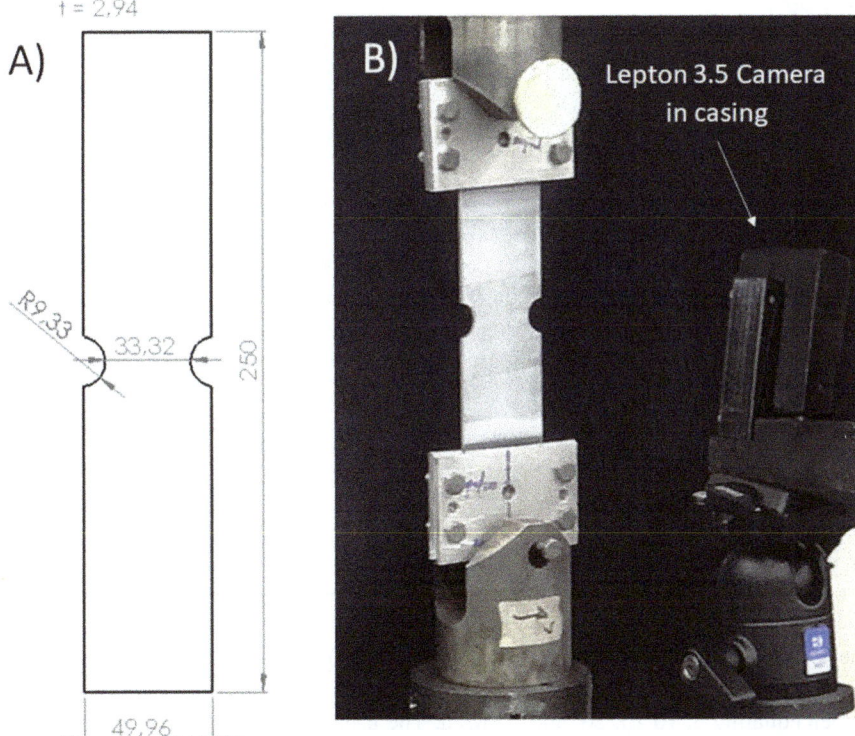

Fig. 4.2 U-notch strip: (**a**) Strip geometry, dimensions in mm; (**b**) Experimental setup

Figure 4.2b presents the experimental setup. The U-notch bar fatigue test was performed with a maximum load of 1.80 kN and a minimum of 0.18 kN, equivalent to a uniaxial nominal-net stress range of 16.5 MPa. This value is above the fatigue limit determined in [20] equal to 13.8 MPa for $N = 10^5$ cycles.

During each cyclic test, the surface temperature of the specimens was recorded in real time by two different thermal cameras: a low-cost FLIR Lepton 3.5 [21] (160 × 120 pixels resolution uncooled microbolometers, 8.7 Hz acquisition rate and <50 mK sensitivity) and a medium-cost FLIR A655sc camera [22] (640 × 480 pixels resolution uncooled microbolometers, up to 50 Hz acquisition rate, 30 mK sensitivity).

For each camera, two different approaches were taken. One approach worked with temperature data analyzed with the Stress Photonics' DeltaTherm II software [23] (using constant amplitude mode and intervals of 2 s for accumulation time and 32 seconds for integration time). The other approach worked with temperature data analyzed with the FLIR's ResearchIR software (for the medium cost camera) or with the Parabilis Thermal Imaging software [24], which used a Raspberry Pi 4 Model B and a PureThermal 2 Smart I/O Module (for the low-cost Lepton 3.5 camera). In this second approach the measured raw temperature data was post-processed using a MATLAB in-house developed algorithm.

Parabilis Thermal Imaging is an open-source software that runs on a computer with a Linux-based operating system, which can be a desktop or, in this case, a Raspberry Pi computer. The Lepton camera was integrated into the Purethermal board adapting I/O camera pins to the USB interface.

The developed MATLAB algorithm used the least square method to fit the measured values of temperature variation (ΔT), which were obtained from the thermal images captured with both thermal cameras. The algorithm was designed to estimate the test frequency (under cyclical loading conditions) from the temperature fitting process. During all tests the Lepton camera was set up with the maximum acquisition rate of 8.7 fps, while the A655sc camera used an acquisition rate of 50 fps for the Deltatherm software approach and 8.7 fps for the MATLAB algorithm approach.

The TSA signal in the in-house algorithm was integrated over 278 frames with an accumulation time of 2 s, equivalent to approximately 32 s of collection time, meaning that a total of 278 frames were collected and then processed with an interval of 2 s to form a single TSA image, the results representing the average of 16 images combined.

The finite element method was used to assess the behavior of the U-notched strip specimen under load. A model consisting of 37,590 elements (SOLID186-SURF154-COMBIN14) and 191,169 nodes, Fig. 4.3, was constructed using a commercial package ANSYS v18.1 to simulate the stress state around the notches and the crack initiation and propagation. The continuous-line true-stress-true strain curve depicted in Fig. 4.1b was inputted in the numerical model. A 1.62 kN uniaxial

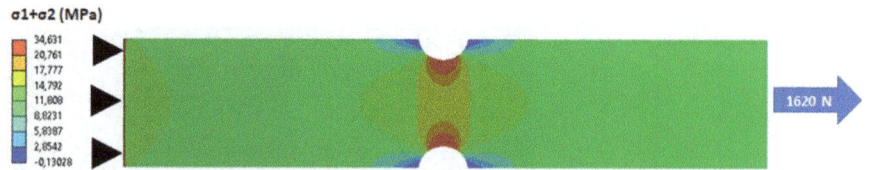

Fig. 4.3 U-notch strip finite element model loaded by a uniaxial 1.62 kN force and resulting invariant (σ1 + σ2) stress distribution

force equivalent was applied in one of the ends while the displacements of the other end were restrained. Moreover, the model was also subjected to a series of loads for validation via DIC comparison.

In addition, prior to the uniaxial tensile test the dog bone specimen was used for a calibration analysis between the measured temperature variations and the cameras response signal.

4.3 TSA Signal Calibration

To correlate the cyclic invariant stress range with temperature range, it is necessary to quantify the gain from the thermoelastic signal [4] by means of a calibration process. For this purpose, several tests were performed on specimens with known stress states ($\Delta\sigma1 + \Delta\sigma2$).

Using the dog bone specimen uniaxially loaded, $\Delta\sigma1 = \Delta F/A$ and $\Delta\sigma2 = 0$, where ΔF is the applied load range and A is the cross section being analyzed. A series of tests was carried out with uniform stress amplitudes actuating in the smallest cross section of specimen equaling 5, 10 15, 20, and 25 MPa. The used load ratio (R) was equal to 0.1.

Figures 4.4 and 4.5 present the post-processed TSA magnitude values for both approaches and both cameras for the tested loads, respectively, for the A655sc and Lepton 3.5 camera. The left images show the magnitude values measured using the in-house algorithm while the right images show the values obtained with the Deltatherm II software. The results for the TSA magnitude values are displayed in °C for the in-house algorithm and in camera units for the Deltatherm II software.

Figure 4.6 shows the relationship between the applied stress amplitude at the smallest cross section of the dog bone specimen and the corresponding TSA magnitude response, the results for in-house algorithm at left and right for the Deltatherm 2 software. The high noise data originated from low stress amplitudes measured with the Lepton camera were disregarded from the analysis.

4.4 Results and Discussion

This section presents results determined with the TSA technique using both cameras compared with DIC and or FE results. The presented results encompass stress distributions, stress concentration factors, and stress intensity factors for cracks that originated from the notches during the cycled tests.

4.4.1 Stress Distribution and Stress Concentration Results

Finite element and DIC techniques provided relevant information about full-field stress and strain distributions actuating in the U-notch strip specimen as presented in Fig. 4.7. It was found a good agreement between displacement and strain results measured by the DIC and determined from the FE model for the positions located at the notches, also known as hot spots. Figure 4.7 shows the principal strain results for a load equal to 1kN.

The TSA magnitude outputs (proportional to the elastic principal stress-invariant ranges) for the U-notch strip using the two cameras and two processing approaches are presented in Fig. 4.8. The hot spots are located at the edge of the notches, positions that experience greater stress ranges and consequentially greater temperature variations.

The relation between the TSA magnitude signal and the principal stress state was made using the linear calibration relationships given by the slopes of the tendency lines depicted in Fig. 4.6. DIC, FEM, and TSA results determined along the meridional line that passes through the center of the specimen, going from one notch to another as shown in Fig. 4.7, are

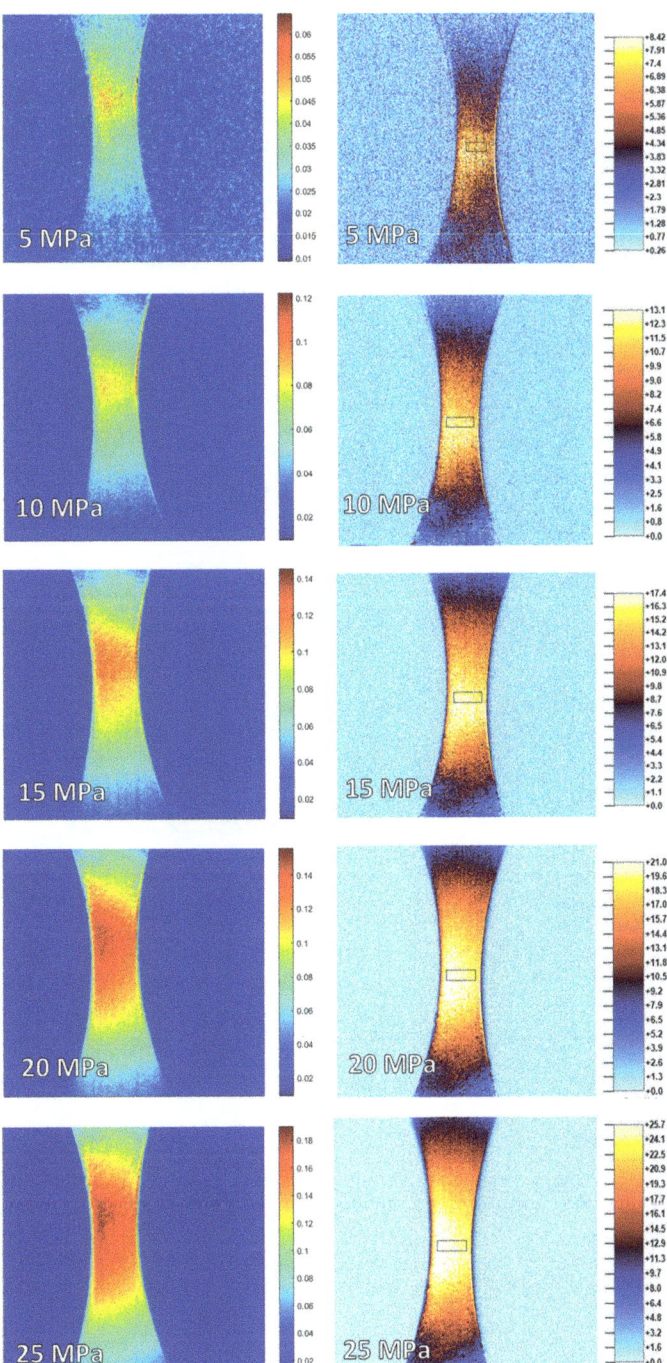

Fig. 4.4 Measure values of TSA magnitude using the FLIR A655sc camera: in-house algorithm (left column), Deltatherm II software (right column)

plotted in Fig. 4.9 in terms of the sum of the principal stress for an alternate load equal to 1.62 kN. Although some discrepancy exists, results showed very good comparison among the TSA, the DIC, and the FE results.

The TSA signal is composed of a magnitude proportional to the temperature variation caused by the thermoelastic effect, and a phase angle, which is related to the signal of the actuating stresses. Vieira et al. [25] studied fatigue crack propagation on a SAE keyhole specimen made of polycarbonate, finding and defining a phase angle modulus equal to 45^0 to differentiate the magnitude signal (tension or compression). This value was found based on the specimen geometry and expected behavior. In

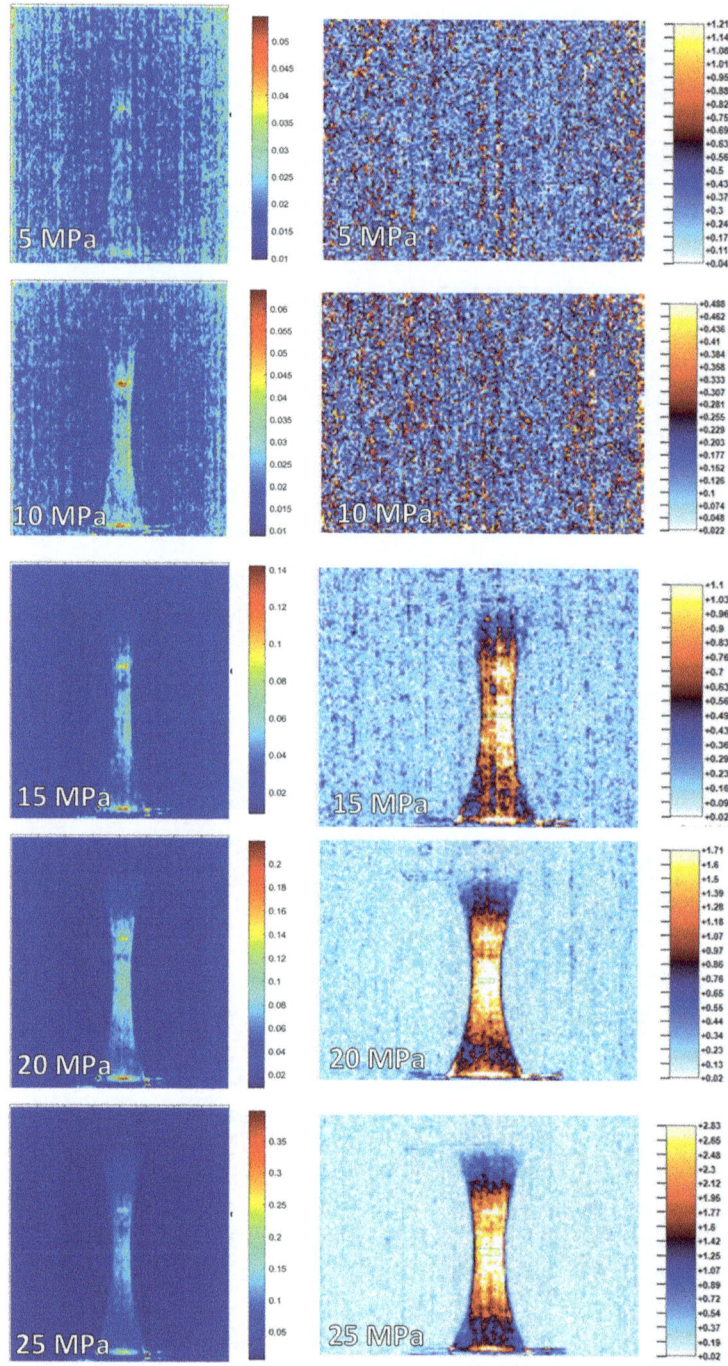

Fig. 4.5 Measure values of TSA magnitude using the FLIR Lepton 3.5 camera: in-house algorithm (left column), Deltatherm II software (right column)

the present paper all TSA measured ranges were positive, the phase angle approaching to zero degrees, in accordance with the determined stress results determined using the DIC and FE techniques.

The stress/strain concentration factors (defined using the gross nominal section) corresponding to both U-notches were measured and estimated using the experimental and numerical techniques. These values are shown in Table 4.2.

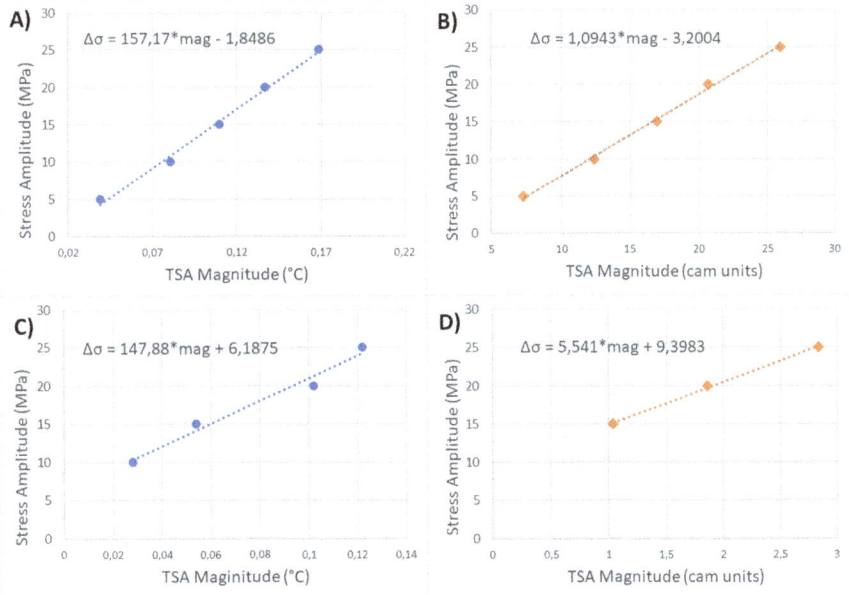

Fig. 4.6 Relationship between applied stress amplitude with TSA magnitude response: (**a**) in-house algorithm FLIR A655sc; (**b**) Deltatherm II FLIR A655sc; (**c**) in-house algorithm FLIR Lepton 3.5; (**d**) Deltatherm II Lepton 3.5

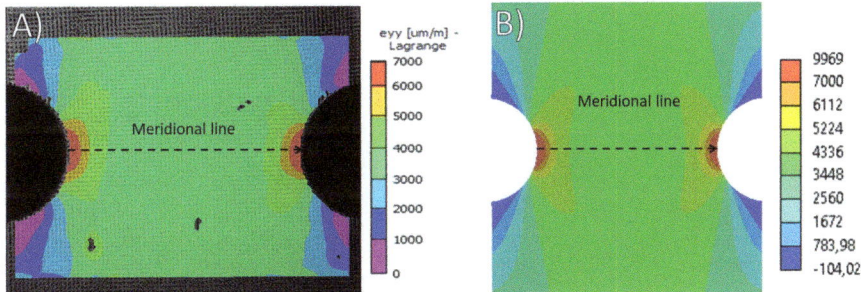

Fig. 4.7 U-notch strip specimen principal strain results for a load equal to 1kN: a) DIC; b) FEM

4.4.2 Stress Intensity Factor Results

While TSA is in essence a linear elastic technique, some plasticity occurs in and around hot spot positions and consequently fatigue cracks may be initiated after a considerable number of load cycles. Due to the nature of the technique, it can be used to locate, monitor the development of a fatigue crack, and furnish quantitative information on the stress intensity factor [25, 26]. Therefore, besides providing information on the stress state of a cyclic loaded sample, TSA can also be extended as a fatigue monitoring tool, supplying real-time information about the specimen damage condition.

The birth and evolution of fatigue crack is accompanied by a local stress relief, which is perceived as the decrease in the stress level compared to an earlier stage. Large cracks can be easily spotted, while small ones need several cycles until can be detected [27]. If the crack is already present in the specimen or the measurement begins after its development, it can be identified by the stress concentration originated at the crack tip and the stress relief along the crack surfaces.

As already noticed using data from the FLIR camera in previous publications [13–15, 27], the present TSA monitoring of the U-notch specimen with the Lepton camera helped to detect an increase in the stress level in the right notch of the specimen, symbolizing the initiation and start of propagation of a fatigue crack. Thereafter, the crack existence becomes noticeable in 856 cycles. Figure 4.10 shows a sequence of images where becomes evident the growth of a crack in the hot-spot area.

Different approaches have been developed to estimate stress intensity factor ranges (ΔK) from the analysis of the TSA response [4, 26, 28]. The data at the tip of the crack and very close to it tend to be very noisy due the presence of plasticity and high stress gradients. Treatment of the TSA data near the crack tip such as suggested by the Stanley's method [30] uses the first two terms of the Westergaard's stress function to adjust the thermoelastic data around the cracks in modes I and II. The

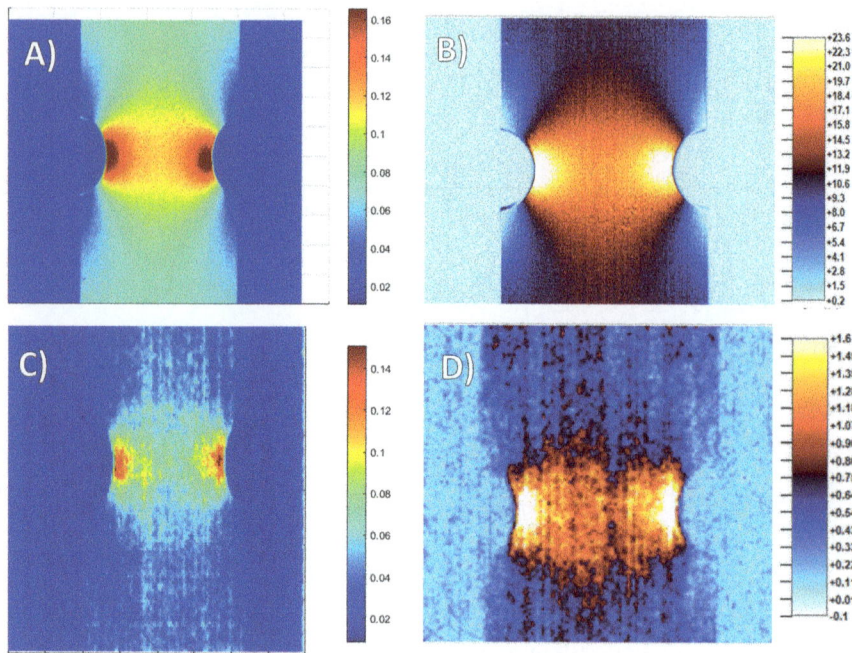

Fig. 4.8 TSA magnitude of the U-notch strip specimen: (**a**) in-house algorithm with FLIR A655sc camera; (**b**) Deltatherm II with FLIR A655sc camera; (**c**) in-house algorithm with FLIR Lepton 3.5 camera; (**d**) Deltatherm II with FLIR Lepton 3.5 camera

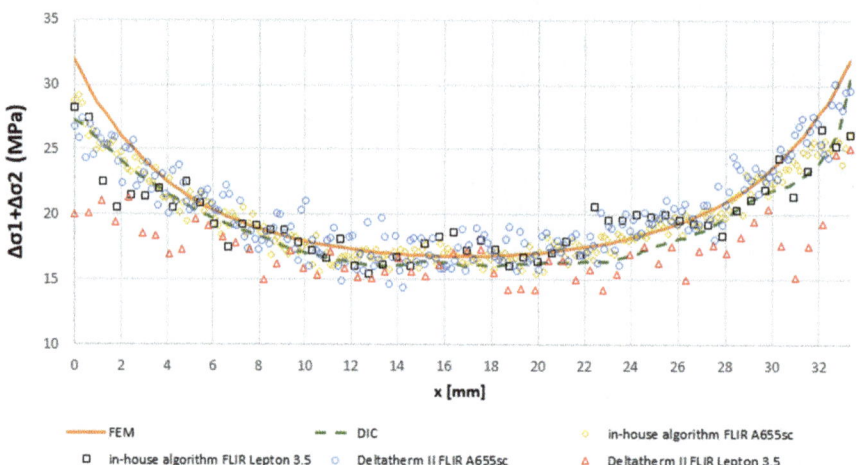

Fig. 4.9 Sum of the principal stress along of the meridional line of the U-notch strip for different techniques

Table 4.2 Stress/strain concentration factors corresponding to both U-notches

Method	Stress/strain concentration factor	
	Left notch	Right notch
FEM	2.87	2.87
DIC	2.88	2.97
In-house algorithm FLIR A655 sc	2.86	2.80
In-house algorithm FLIR lepton 3.5	2.85	2.86
Deltatherm II FLIR A655 sc	2.85	2.94
Deltatherm II FLIR lepton 3.5	2.52	2.85
Analytical Formulation [1]	2.95	2.95

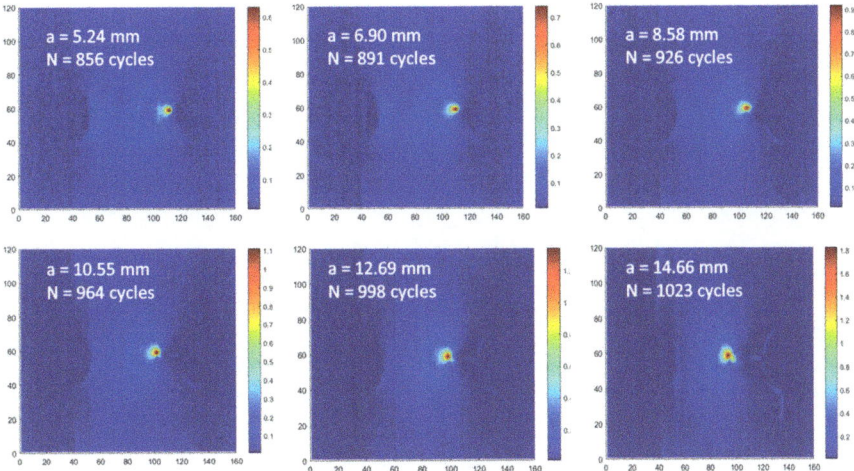

Fig. 4.10 Fatigue crack propagation on the right notch of the U-notch strip

Fig. 4.11 Relationship between the vertical distance from the crack tip and ($1/S^2$) employed in the methodology of Stanley for the calculation of ΔK using TSA for the U-notch specimen with a 12.69 mm crack

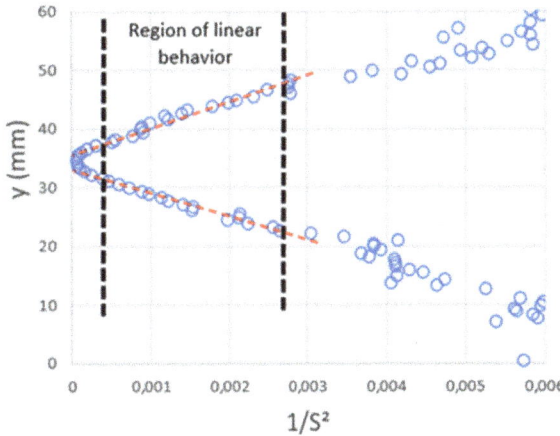

method is only valid for the region of elastic and linear behavior near the crack tip and uses the gradient of this region to calculate ΔK. This method was applied herein.

Figure 4.11 presents points given by pair of coordinates where the horizontal axis' values are calculated using the inverse square of the TSA signal ($1/S^2$) plotted and the ordinate values are given by the vertical distance (y) to the crack. The Stanley's method uses Eq. (4.1) to determine ΔK as a function of the slope of the of the dashed line presented in Fig. 4.11.

$$\Delta K_{TSA} = \sqrt{4\pi \cdot A^2 \cdot \tan\, \alpha / \left(3\sqrt{3}\right)} \tag{4.1}$$

where $\tan \alpha$ is the slope of the y distance versus $1/S^2$ and A is the material calibration constant.

A crack was propagated in Mode I along the meridional line of the tested specimen, starting from the right U-notch. Crack lengths were measured with a caliper while TSA data were being collected. With a spatial resolution of 0.55 mm per pixel from the Lepton camera, stress intensity factors in mode I were calculated using the TSA-Stanley's approach and are compared with FEM determined ΔK results, which were based on an approach described in reference [29]. Results from this comparison are presented in Fig. 4.12.

Figure 4.13 presents plots of crack growth rate against the actuating stress intensity factor in mode I calculated using the TSA signal measured with the Lepton camera. These experimentally determined results were compared with the calculated FEM numerical results and show good agreement.

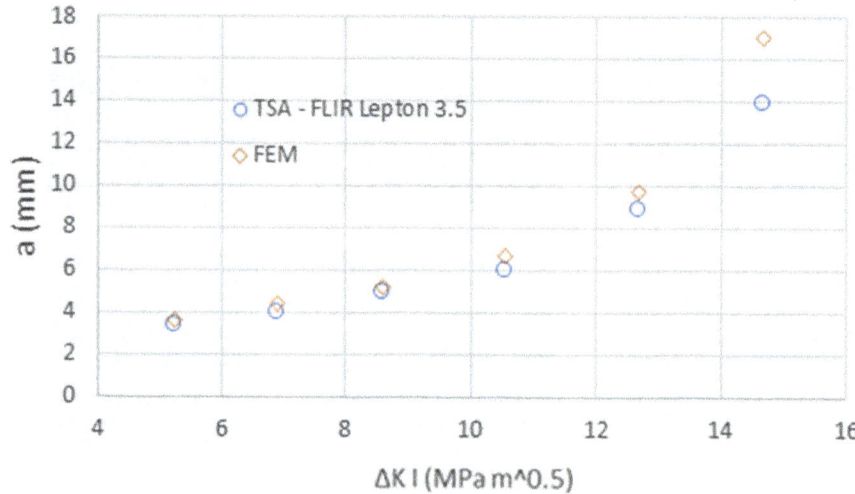

Fig. 4.12 Stress intensity factors in mode I for various crack lengths calculated using FEM and the Stanley's method based on the TSA signal measured with the FLIR Lepton camera

Fig. 4.13 Stress intensity factors in mode I plot against the crack growth rate using TSA signal measured with the Lepton camera and FEM

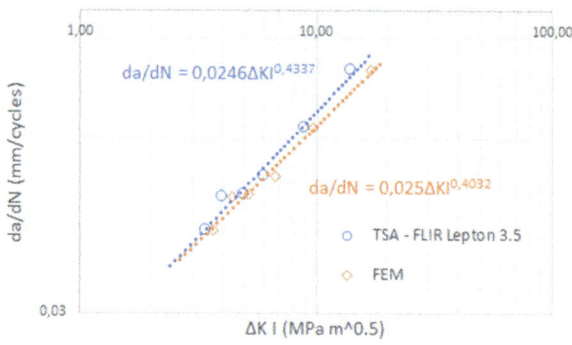

4.5 Conclusions

It has been showed along the years that TSA is a powerful tool for monitoring and evaluating structural health. Further advances in thermal camera technology have enabled much less inexpensive models, such as FLIR Lepton 3.5, to emerge, making it more accessible to be used in stress analysis. Although the quality of the results may be seen proportional to the investment in the camera, such low-cost cameras may be used as efficient alternatives for stress and damage monitoring in structures, mainly when large number of points need to be monitored considering affordable cost restrictions.

The present paper combines a set of different experimental and numerical techniques to assess the stress distributions and monitor fatigue damage of a U-notch specimen made of polycarbonate. DIC and FEM were used to provide relevant data about the stress/strain state at the notches area. Furthermore, the TSA technique was applied using a low-cost Lepton camera and a medium-cost FLIR A655sc camera. Data analysis used two different approaches: an in-house developed MATLAB algorithm and the commercial software Deltatherm II.

The results obtained using the Lepton camera were similar to those measured and obtained in other ways. The determined stress concentration factors at the hot spots presented good agreement among the applied experimental and numerical techniques. Another two objectives were achieved by showing that crack initiation and crack growth could be monitored using the low-cost camera as well as that stress intensity factors could also be accurately measured, therefore anticipating promissory applications of TSA in the integrity assessment of actual structures and their components.

References

1. Peterson, R.E., Plunkett, R.: Stress concentration factors. J. Appl. Mech. **42**(1), 248 (1975)
2. Anderson, T.L.: Fracture Mechanics: Fundamentals and Applications. CRC Press, New York (2017)
3. Castro, J.T., Meggiolaro, M.A.: Fadiga - Técnicas e Práticas de Dimensionamento Estrutural sob Cargas Reais de Serviço, 2nd edn. CreateSpace, Scotts Valley (2009)
4. Greene, R.J., Patterson, E.A., Rowlands, R.E.: Thermoelastic stress analysis. In: Sharpe, W.N. (ed.) Springer Handbook of Experimental Solid Mechanics, pp. 743–767. Springer, Berlin (2008)
5. Sirca, G.F., Adeli, H.: Infrared thermography for detecting defects in concrete structures. J. Civ. Eng. Manag. **24**, 508–515 (2018)
6. Avdelidis, N.P., Hawtin, B.C., Almond, D.P.: Transient thermography in the assessment of defects of aircraft composites. NDT & E Int. **36**, 6 (2003)
7. Sakagami, T., Komiyama, T.: Thermographic nondestructive testing for concrete structures. J JSNDI. **47-10**, 723–727 (1998)
8. Sakagami, T., Kubo, S.: Development of a new crack identification method based on singular current field using differential thermography. SPIE Proc. **3700**, 369–376 (1999)
9. Sakagami, T., Kubo, S., Teshima, Y.: Fatigue crack identification using near-tip singular temperature field measured by lock-in thermography. Proc SPIE. **4020**, 174–181 (1999)
10. Sakagami, T., Nishimura, T., Kubo, S.: Development of a self-reference lock-in thermography and its application to crack monitoring. Proc. SPIE. **5782**, 379–387 (2005)
11. Sakagami, T.: Remote nondestructive evaluation technique using infrared thermography for fatigue cracks in steel bridges. Fatigue Fract. Eng. Mater. Struct. **38**, 755–779 (2015)
12. Sakagami, T., Mizokami, Y., Shiozawa, D., Fujimoto, T., Izumi, Y., Hanai, T., Moriyama, A.: Verification of the repair effect for fatigue cracks in members of steel bridges based on thermoelastic stress measurement. Eng Fracture Mech. **183**, 1–12 (2017)
13. Paiva, V. Maneschy, R.J., Freire, J.L., Gonzáles, G.L.G., Vieira, R.D., Vieira, R.B.: Fatigue monitoring of a dented piping specimen using infrared thermography. PVP 2018, paper no. PVP 2018-84597, ASME pressure vessels & piping conference, Praga, Czech Republic. Proceedings of the PVP 2018, paper no. PVP 2018-84597, p. 1-7 (2018)
14. Paiva, V., Maneschy, R. J., Freire, J.L., Gonzáles, G. L. G. Vieira, R. D. Ribeiro, A.S. Almeida: Fatigue assessment and monitoring of a dented pipeline specimen, PVP 2019, paper no. PVP 2019-93663, ASME pressure vessels & piping conference, 2019, San Antonio, TX, USA. Proceedings of the PVP 2019, paper no. PVP 2019-93663, ASME pressure Vessels & Piping Conference, p. 1–8 (2019)
15. Paiva, V., Maneschy, R.J., Freire, J.L., Gonzáles, G.L. Vieira, G., Ribeiro, R.D., Almeida, A.S., Diniz, J.L.C: Fatigue monitoring of a dented pipeline specimen using infrared thermography, DIC and fiber optic strain gages, SEM conference and exposition on experimental and applied mechanics, Reno NV. Proceedings of the SEM Conference and Exposition on Experimental and Applied Mechanics, p. 1–10 (2019)
16. Paiva, V.E., Vieira, R.D., Freire, J.L.F: Fatigue properties assessment of API 5L Gr. B pipeline steel using infrared thermography. In: SEM Conference and Exposition on Experimental and Applied Mechanics, 2018, Greenville SC. Proceedings of the SEM Conference and Exposition on Experimental and Applied Mechanics. v. 1. p. 1–7 (2018)
17. Paiva, V.E.L.; Freire, J.L., Etchebehere, R.C.: Assessment of chain links using infrared thermography. In: CONAEND & IEV 2018–378, Congresso Nacional de Ensaios Não Destrutivos e Inspeção, 21 IEV Conferência Internacional sobre Evaluación de Integridad y Extensión de Vida de Equipos Industriales ABENDI, 2018, São Paulo. Anais do Conaend& IEV 2018, pp. 1–17 (2018)
18. Rajic, N., Rowlands, D.: Thermoelastic stress analysis with a compact low-cost microbolometer system. Quant Infrared Thermogr J. **10**(2), 135–158 (2013)
19. Weihrauch, M., Middleton, C., Greene, R., Patterson, E.: Low-cost thermoelastic stress analysis. In: Residual Stress, Thermomechanics & Infrared Imaging and Inverse Problems, pp. 15–19. Springer, Cham (2020)
20. Vieira. R.B.: Thermography Applied to the Study of Fatigue in Polycarbonate. Masters dissertation. PUC-Rio (2016)
21. https://www.flir.com/globalassets/imported-assets/document/flir-lepton-engineering-datasheet.pdf. Accessed 19 Feb 2020
22. https://www.flirmedia.com/MMC/THG/Brochures/RND_011/RND_011_US.pdf. Accessed 19 Feb 2020
23. Stress Photonics Inc.: DeltaTherm 2 Manual, v6(2016)
24. https://github.com/Kheirlb/purethermal1-uvc-capture. Accessed 21 Feb 2020
25. Vieira, R.B., Gonzáles, G.G., Freire, J.L.F.: Thermography applied to the study of fatigue crack propagation in polycarbonate. Exp. Mech. **58**(2), 269–282 (2018)
26. Díaz, F.A., Patterson, E.A., Yates, J.R.: Application of thermoelastic stress analysis for the experimental evaluation of the effective stress intensity factor. Frattura ed Integrità Strutturale. **7**(25), 109–116 (2013)
27. Paiva, V.E.L.: Modern Experimental Techniques with an Emphasis on Infrared thermography to the Assessment of Fatigue Components with Dents. PhD Thesis. PUC-Rio (2020)
28. Marsavina, L., Tomlinson, R.A.: A review of using thermoelasticity for structural integrity assessment. Fracture Struct Integr: Annals. **2014**, 8 (2014)
29. Stanley, P., Chan, W.K.: The determination of stress intensity factors and crack tip velocities from thermoelastic infra-red emissions. In: Proceedings of International Conference of Fatigue of Engineering Materials and Structures, pp. 105–114. IMechE, Sheffield (1986)

Chapter 5
Evaluation of Thermal Deformation of Fastening Structure of Carbon Fiber Reinforced Plastic and Aluminum Using Stereo Image Correlation

Ayumu Fuchigami, Satoru Yoneyama, Ken Goto, and Kosei Ishimura

Abstract This paper describes the evaluation of thermal deformation of the CFRP-aluminum fastening structure using stereo DIC (stereo digital image correlation). This structure is used at the FOB (Fixed Optical Bench) of the astronomical satellite "ASTRO-H." The fastening structure is heated up with a self-made thermostat, and the in-plane displacements on the three-dimensional surface of it are measured using stereo DIC. Then, the strains are evaluated and the results are compared with those obtained by FEM. By comparing experimental results with those of finite element analysis, the finite element model is improved, then analysis of thermal deformation of the fastening structure is made more accurate and easily.

Keywords CFRP · Aluminum alloy · Thermal deformation · Fastening structure · Stereo digital image correlation

5.1 Introduction

Space structures with observations such as artificial satellites require a highly accurate and large-scale structural system in order to suppress alignment errors in the orbit of observation equipment. Since space structures are often large, even minute errors can appear as serious measurement errors when measuring and observing objects, so high accuracy is required. However, these instruments are very costly to experiment on an actual scale; the validity of the analysis model is important. Alignment error factors after launch of thermal deformation are more dominant because artificial satellites orbit the near earth orbit and repeat shade and sunny.

The X-ray astronomical satellite "ASTRO-H" developed by JAXA is an artificial satellite that observes high-temperature, high-energy celestial bodies using X-ray telescopes and so on. Since the soft X-ray telescope mounted on this satellite has a long focal length of 5.6 m, the telescope and the X-ray imaging detector must be installed separately. To achieve this, FOB (Fixed Optical Bench) is installed. FOB is a truss structure that combines three plates, top, middle, and lower, with tubes made of CFRP [1, 2]. The three plates and tubes are connected by a member called a tube fitting made of aluminum alloy (Fig. 5.1). In order to prevent the FOB from expanding and contracting in the longitudinal direction and falling in the lateral direction due to heat input in the orbit, the CFRP of the tube has a devised fiber orientation so that the coefficient of thermal expansion is negative, by offsetting with the positive coefficient of thermal expansion of the tube fitting, the longitudinal direction of the FOB as a whole falls within the coefficient of thermal expansion of about 0.1 ppm or less. However, in a comparison between the thermal deformation measurement results and the analysis results of ASTRO-H conducted in 2012, a significant difference was found in the translational displacement of the top plate. The analysis error of the top plate on which the telescope is mounted causes misalignment of the optical axis of the telescope and the detector. Then, the cause was investigated for improvement, the analysis error of the truss part of the FOB that supports the top plate was considered. More detailed correlation of the model was needed.

A. Fuchigami (✉) · S. Yoneyama
Aoyama Gakuin University, Tokyo, Japan
e-mail: c5620147@aoyama.jp; yoneyama@me.aoyama.ac.jp

K. Goto
Japan Aerospace Exploration Agency, Tokyo, Japan
e-mail: goto.ken@jaxa.jp

K. Ishimura
Waseda University, Tokyo, Japan
e-mail: ishimura@waseda.jp

© The Society for Experimental Mechanics, Inc. 2022
S. L. B. Kramer et al. (eds.), *Thermomechanics & Infrared Imaging, Inverse Problem Methodologies, Mechanics of Additive & Advanced Manufactured Materials, and Advancements in Optical Methods & Digital Image Correlation, Volume 4*, Conference Proceedings of the Society for Experimental Mechanics Series, https://doi.org/10.1007/978-3-030-86745-4_5

Fig. 5.1 Fixed optical
bench [1]

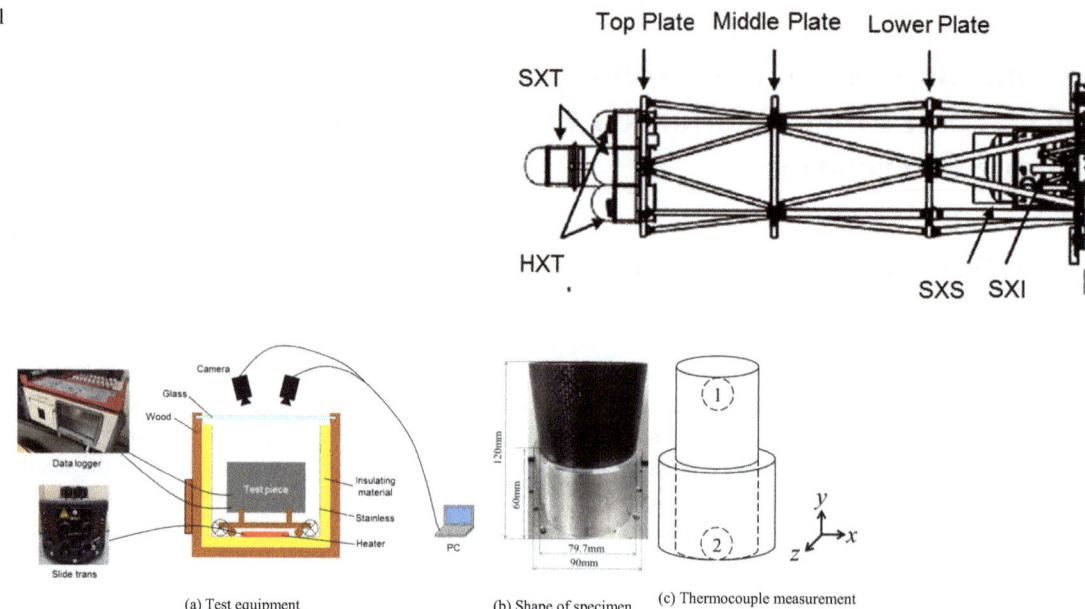

Fig. 5.1 Fixed optical bench [1]

(a) Test equipment   (b) Shape of specimen   (c) Thermocouple measurement positions

Fig. 5.2 Thermal deformation measurement test: (a) test equipment, (b) shape of specimen, (c) thermocouple measurement positions

Therefore, in this study, we focus more on the fastening part, by using the stereo image correlation method [3, 4], which enables non-contact and full-field measurement, thermal strain measurement on the three-dimensional surface of the fastening body of CFRP and aluminum alloy is performed. We aim to establish a thermal deformation analysis method for the fastening structure by correlating the model by comparing the experimental results with the finite element analysis.

5.2 Thermal Deformation Measurement Test Device

First, a thermostatic chamber is prepared for thermal deformation measurement. Figure 5.2a shows a schematic drawing of the thermostatic chamber and the experimental equipment. A silicon rubber heater is used for heating, and the temperature is controlled by voltage control using the slide trans RSA-10 of Tokyo Rikosha Co., Ltd. In addition, two fans are installed on the bottom of the thermostatic chamber to circulate the air and remove thermal fluctuations. The test piece is placed on a wooden table that stands directly above the heater. Furthermore, the surface temperature of the test piece is obtained by attaching a K-type thermocouple with Kempton tape and using the data logger TDS-530 manufactured by Tokyo Sokki Co., Ltd. In addition, ISTRA-4D manufactured by DANTEC DYNAMICS is used for displacement measurement.

5.3 Thermal Deformation Measurement Test Method

The thermal deformation of the fastening structure test piece was measured, and the thermal deformation behavior of the fastening structure was investigated. As shown in Fig. 5.2b, the test piece consists of CFRP and A5056. CFRP is made by laminating 17 plain weave prepregs at 0/90 and molding at 130 °C. It has a cylindrical shape with a length of 120 mm, an outer diameter of 79.7 mm, and a thickness of 3 mm. The A5056 is a cylindrical one with a length of 60 mm, an outer diameter of 90 mm, and a thickness of 5 mm. They bonded with an epoxy adhesive Hysol EA 9396 and cured at room temperature. Therefore, the CFRP test piece is wrapped in A5056. The temperature condition was heating from room temperature to around 90 °C. Figure 5.2c shows the position where the thermocouple is attached, and the dotted line shows that it is inside the test piece.

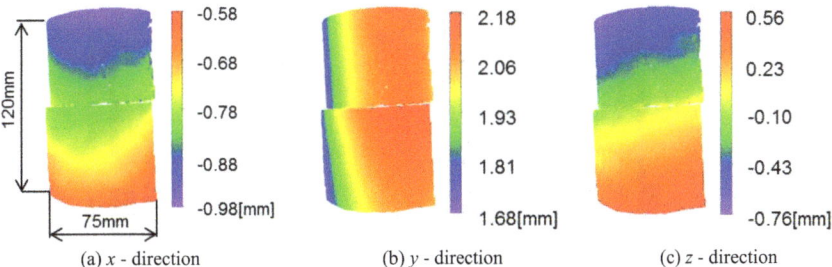

Fig. 5.3 Displacement distribution: (**a**) x–direction, (**b**) y–direction, (**c**) z–direction

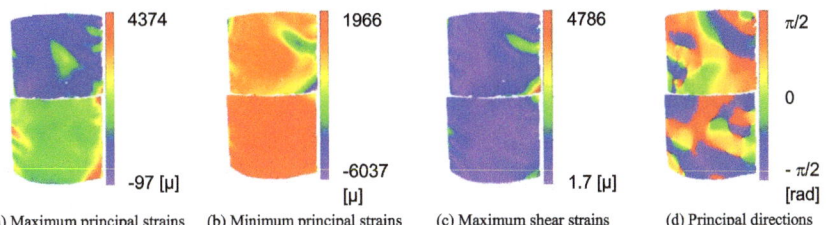

Fig. 5.4 Strain distribution: (**a**) maximum principal strains, (**b**) minimum principal strains, (**c**) maximum shear strains, (d) principal directions

5.4 Thermal Deformation Measurement Result

Figure 5.3 shows (a) x-direction displacements, (b) y-direction displacements, (cc) z-direction displacements, Fig. 5.4 shows (a) maximum principal strains, (b) minimum principal strains, (c) maximum shear strains, (d) principal directions. The temperature difference before and after deformation was 67.1 °C.

From Fig. 5.3, the displacement in the z direction increases toward the bottom of the test piece; the difference in the amount of deformation between CFRP and the aluminum alloy is noticeable. On the other hand, the amount of CFRP deformation is larger for the displacement in the x direction. Regarding the displacement in the y direction, there is no difference in the amount of deformation depending on the material, and the entire test piece is deformed uniformly. But since the displacement on the right side of the contour diagram is larger, it can be seen from the results in the x and y directions that the entire test piece is tilted. Furthermore, the y-direction displacement has the largest amount of deformation; it is considered that the y-direction deformation, that is, the axial deformation, is the most sensitive to the behavior of FOB. Next, from Fig. 5.4, the principal strain distribution shows some variations due to displacement measurement errors, but it is almost uniform in many parts. Regarding (a) maximum principal strain and (b) minimum principal strain, it can be seen that the strain distributions are clearly separated between the CFRP portion and the aluminum alloy portion. Next, (c) the maximum shear strain was initially thought to appear near the joint due to the difference in the amount of deformation of the two materials, but it did not appear in this test. The cause of this is thought to be the behavior of the adhesive. I would like to mention this in the future through comparison with analysis.

5.5 FEM Analysis of Fastening Structure

Next, in order to compare with the thermal deformation measurement, FEM analysis of the fastening structure was performed. The nonlinear general-purpose structural analysis program MARC/Mentat was used for the analysis. The analysis model is a 4-node quadrilateral element with 344 nodes and 300 elements, and is an axisymmetric solid model around the y - axis. In addition, the y-direction constraint shown in the figure is included as a constraint condition. The CFRP part was 120 mm long and 3 mm thick, the adhesive part was 60 mm long and 0.15 mm thick, and the A5056 part was 60 mm long and 5 mm thick. The temperature conditions are the same as in the experiments. A5056 and adhesives are defined as isotropic, and CFRP is defined as orthogonally anisotropic materials. The in-plane characteristics are provided by the ACM Engineering Department of Toray Industries, Inc., but the out-of-plane characteristics shown with asterisk (*) are the values estimated by the specimen provider because there are no actual measurement results.

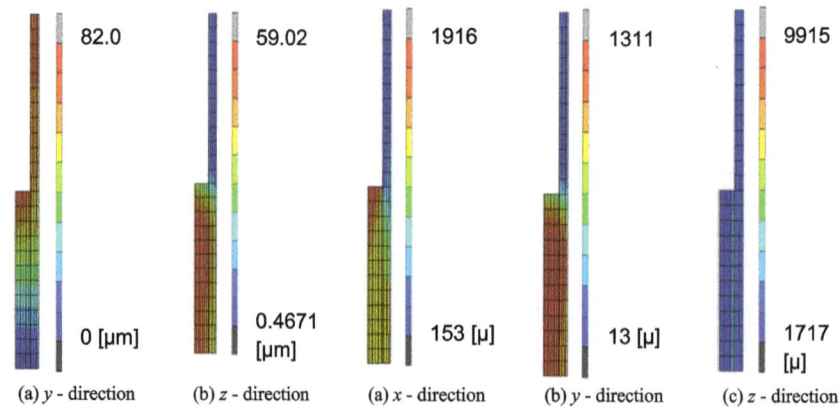

Fig. 5.5 Distribution by FE: (**a**) *y*–direction, (**b**) *z*–direction, (**c**) *x*–direction, (**d**) *y*–direction, (**e**) *z*-direction

Figure 5.5 shows the analysis results of (a) *y*-direction displacement and (b) *z*-direction displacement, (c) *x*-direction strain, (d) *y*-direction strain, and (e) *z*-direction strain. Since this result does not consider the temperature dependence of the material properties, it is considered that the analysis is not so accurate. Therefore, in the future, after measuring the material properties of each material and inputting the results, we plan to pick out the strain value of the part corresponding to the imaging surface of the actual thermal deformation measurement test and compare it with the actual experiment.

5.6 Conclusion

We aim to establish a thermal deformation analysis method for the fastening structure by correlating the model by comparing the experimental results with the finite element analysis. The thermal deformation of the fastening structure was measured, from the displacement distributions, it was shown that the entire test piece behaved as if it were tilted, and that the FOB behavior was highly sensitive to axial deformation. In addition, the strain distribution showed that the test piece was deformed almost uniformly, suggesting that the adhesive might have an effect on the fact that no shear strain appeared at the joint. We established an FEM analysis method for thermal deformation of the fastening structure using an axisymmetric model and examined a comparison method with the experimental results. In the future, we will conduct tensile tests and TMA tests of adhesives with the aim of analyzing the temperature dependence of each material property. After inputting those measurement results into FEM, model correlation will be performed.

References

1. Awaki, H., Ishida, M., Okajima, T.: X-ray optical system of "Hitomi" Satellite. Monthly Rep Astron. **112**, 460–470 (2019)
2. Ishimura, K., Minesugi, K., Kawano, T., Wada, A., Ishida, M., Natsukari, C., Shoji, K., Tsushima, M., Ikeda, M., Omagari, K., Kumashita, K., Tachikawa, K., Abe, K., Kito, A., Iizuka, R.: Thermal deformation test of ASTRO-H precise large space structures. Lect Space Sci Technol Assoc. **2012**, 56 (2012)
3. Yoneyama, S.: Basic principle of image correlation method and in-plane displacement and strain distribution measurement procedure. Japan Compos Mater Soc. **40**, 135–145 (2014)
4. Yoneyama, S.: Displacement and strain measurement method for 2D and 3D surfaces using image correlation method. Japan Soc Plastic Processi. **55**, 979–983 (2014)

Chapter 6
Full-Field Analysis of the Strain-Induced Crystallization in Natural Rubber

S. Charlès and J. -B. Le Cam

Abstract The strain-induced crystallization is generally considered to be responsible for the excellent properties of natural rubber, especially its remarkable crack growth resistance.

The crystallinity of rubber is classically studied by using X-ray diffraction (XRD). The XRD technique gives access to the crystallinity but also information of paramount importance on the crystalline phase structure (Rajkumar, et al., Macromolecules 39:7004, 2006), chain orientation (Toki, et al., Rubber Chemistry and Technology 42:956-964, 2004), kinetics of crystallization (Trabelsi, Macromolecules 36:9093–9099, 2003), non-exhaustively. However, this method provides this information at one point.

Recently, a new method has been proposed in (Le Cam, Strain 5:54, 2003) for determining the crystallinity, from infrared thermography-based surface calorimetry. In the case of heterogeneous heat source field and large deformations, the method requires combining digital image correlation and infrared thermography. In the present study, this methodology is applied to measure full heat source field in a stretched rubber specimen. The crystallinity as well as its spatial distribution is characterized.

Keywords Strain-induced crystallization · Natural rubber · Infrared thermography · Surface calorimetry · Digital image correlation

6.1 Introduction

Strain-induced crystallization (SIC) is classically mentioned to explain the remarkable properties of natural rubber compared to non-crystallizing elastomers [9–11]. Since its discovery by Katz in 1925 [12], many experimental methods have been used to explore more deeply this phenomenon, such as stress relaxation [13], Raman Spectroscopy [14], birefringence [15], or nuclear magnetic resonance [16]. However, X-ray diffraction remained the most used method to investigate SIC.

While this method gives access to the crystallinity, it also provides information on crystallites morphology and the crystallographic structure. Nevertheless, this kind of information is provided at a single point. This limitation can induce issues in case of crystallinity heterogeneity, such as at a crack tip [17].

Since the SIC phenomena is a strong exothermal phenomenon [18], a new method has been proposed in [8] and validated in [19].

In this study, this method is used to measure crystallinity at several materials points simultaneously in an unfilled natural rubber specimen submitted to a uniaxial tensile loading.

6.2 Experimental Setup

In the present study, an unfilled natural rubber is considered. The specimen geometry corresponds to a thin rectangular specimen, as shown in Fig. 6.1.

S. Charlès (✉) · J. -B. Le Cam
IPR (Institut de Physique de Rennes) – UMR 6251, Univ Rennes, CNRS, Rennes, France
e-mail: sylvain.charles@univ-rennes1.fr

© The Society for Experimental Mechanics, Inc. 2022
S. L. B. Kramer et al. (eds.), *Thermomechanics & Infrared Imaging, Inverse Problem Methodologies, Mechanics of Additive & Advanced Manufactured Materials, and Advancements in Optical Methods & Digital Image Correlation, Volume 4*, Conference Proceedings of the Society for Experimental Mechanics Series, https://doi.org/10.1007/978-3-030-86745-4_6

Fig. 6.1 Specimen
geometry

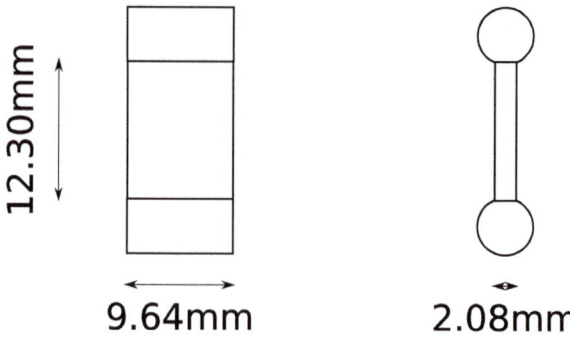

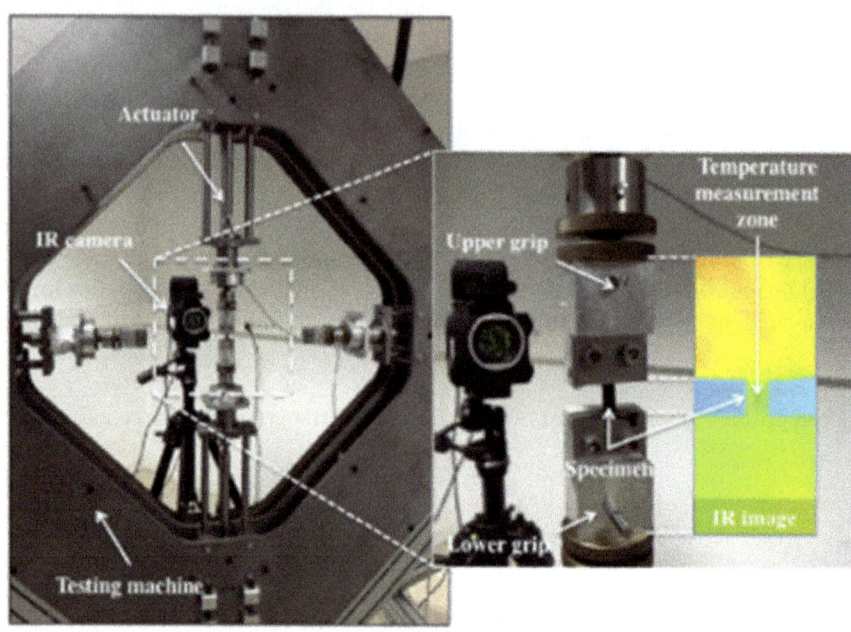

Fig. 6.2 Experimental setup

Figure 6.2 presents an overview of the experimental setup used for this study. The mechanical test is performed using a home-made biaxial testing machine. This machine is composed of four independent RCP4-RA6C-I-56P-4-300-P3-M (IAI) electrical actuators. They are driven by a CON-CA-56P-I-PLP-2-0 controller and four PCON-CA (IAI) position controllers. An in-house LabVIEW program piloted these actuators. This machine enables us to stretch symmetrically the specimen. Therefore, the specimen center is motionless.

The mechanical loading applied consists in three loading/unloading uniaxial cycles. The maximal displacement and loading rate of each actuator are set at 36 mm and 150 mm/min, respectively, which corresponds to a maximal stretch of about 7.

Temperature measurements are performed by using a FLIR infrared camera equipped with a focal array of 640x512 pixels and detectors operating in wavelengths between 1.5 and 5.1 μm. The integration time is equal to 2700 μs, and the acquisition frequency is equal to 20 Hz. The calibration of camera detectors is performed with black body using a one-point NUC procedure at this acquisition frequency. The thermal resolution, or noise equivalent temperature difference (NETD), is equal to 20 mK for a temperature range between 5 and 40 °C. The spatial resolution of the thermal field is equal to 300 μm/px. The infrared camera is switched on several hours before the test in order to stabilize its internal temperature. To deduce the temperature field from R thermography, the emissivity of the material must be known. In the present study, a value of 0.94 has been taken.

6.3   Motion Compensation Technique

Since the specimen undergoes large deformations during the test, material points move from pixel to pixel. Determining the heat source variation and thus the crystallinity at any point requires knowing the temperature variation at any point and any time.

For that purpose, a motion compensation technique has to be used. In the present study, the temperature measurement is performed at the median section of the specimen. The strain field is assumed to be homogeneous, which means that no displacement occurs in the stretch direction in this zone, only displacement along the width due to the cross-section contraction. An algorithm has been developed to follow the borders of the specimen in the IR images and therefore to evaluate the displacement of the points due to the cross-section contraction.

6.4   Crystallinity Measurement

Evaluating the crystallinity requires several steps. Once the motion compensation method has been performed, and the temperature evolution obtained, the heat diffusion equation is first applied. For that purpose, several assumptions are done; they are recalled in [8].

The total heat source produced by the material during the test is expressed as follows:

$$S = \rho C_p \left(\dot{\theta} + \frac{\theta}{\tau} \right) - k \Delta_{2D} \theta$$

with S the heat source produced by the material, θ the temperature variation, ρ the mass density of the material at its undeformed state, C_p the calorific capacity of the material, and k the thermal conductivity of the material.

The second step consists in predicting the heat source S_{el} if no crystallization occurs by using a polynomial form of the heat source. This is possible by considering the heat source variations before the crystallization starts.

$$S_{el} = C_1(I_1 - 3) + C_2(I_1 - 3)^2 + C_3(I_1 - 3)^3$$

with I_1 the first invariant of the Cauchy-Green tensor and C_1, C_2, and C_3 are three parameters to be identified. Figure 6.3 represents both the heat source measured and the heat source predicted using the polynomial form at a single point. The two vertical lines delimit the time range (or the strain range) where the identification was performed.

The difference between the predicted and measured curves at high strains is due to the heat produced by the crystallization phenomenon. Once the heat source due to SIC is identified, the crystallinity is obtained by integrating the heat source due to SIC over the time:

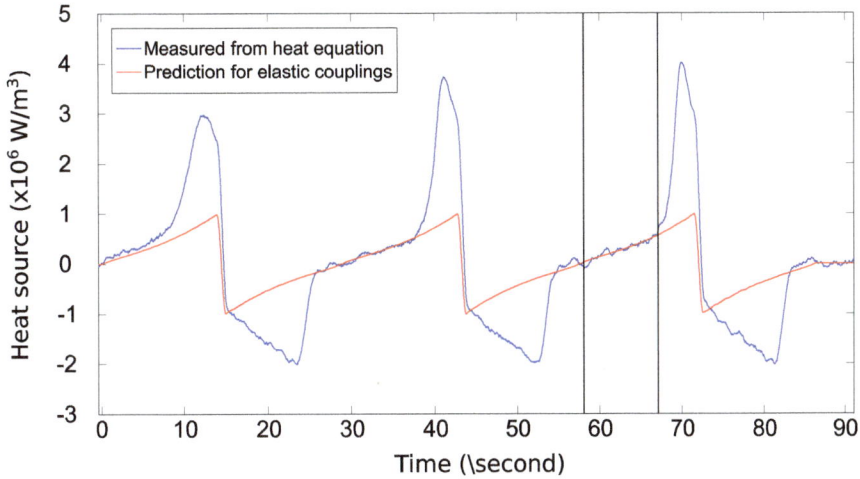

Fig. 6.3 Heat source during the full mechanical test

$$\chi = \int_0^t \frac{S_{\text{cryst}}}{\Delta H} \, dt$$

with χ the crystallinity level, S_{cryst} the heat source due to SIC phenomenon and ΔH the fusion enthalpy.

6.5 Highlighting Crystallinity Heterogeneities

Results will be more precisely detailed and analyzed during the presentation, especially the heterogeneity detected in terms of crystallinity from one point to another.

Acknowledgments The authors thank the National Center for Scientific Research (MRCT-CNRS and MI-CNRS), Rennes Metropole and Region Bretagne for financially supporting this work and the Michelin Company for providing the specimens. Authors also thank Dr. Mathieu Mirroir, Mr. Vincent Burgaud, and Mr. MickaelLefur for having designed the biaxial tensile machine.

References

1. Bunn, C.W.: Proc. R. Soc. London, Ser. **180**, 40–66 (1942)
2. Takahashi, Y., Kumano, T.: Macromolecules. **37**, 4860 (2004)
3. Immirzi, A., Tedesco, C., Monaco, G., Tonelli, A.E.: Macromolecules. **38**, 1223 (2005)
4. Rajkumar, G., Squire, J.M., Arnott, S.: Macromolecules. **39**, 7004 (2006)
5. Toki, S., Sics, I., Ran, S.F., Liu, L.Z., Hsiao, B.S., Murakami, S., Tosaka, M., Kohjiya, S., Poompradub, S., Ikeda, Y., Tsou, A.H.: Rubber Chem. Technol. **42**, 956–964 (2004)
6. Toki, S., Fujimaki, T., Okuyama, M.: Polymer. **41**, 5423–5429 (2000)
7. Trabelsi, S., Albouy, P.-A., Rault, J.: Macromolecules. **36**, 9093–9099 (2003)
8. Le Cam, J.-B.: Strain. **5**, 54 (2018)
9. Trabelsi, S., Albouy, P.-A., Rault, J.: Macromolecules. **36**, 7624–7639 (2003)
10. Hamed, G.R., Kim, H.J., Gent, A.N.: Rubber Chem. Technol. **69**, 807–818 (1996)
11. Lake, G.J.: Rubber Chem. Technol. **68**, 435–460 (1995)
12. Katz, J.R.: Naturwissenschauften. **13**, 410–416 (1925)
13. Gent, A.N.: Trans. Faraday Soc. **28**, 625–628 (1954)
14. Healey, A.M., Hendra, P.J., West, Y.D.: Polymer. **37**, 4009–4024 (1996)
15. Treloar, L.R.G.: Trans. Faraday Soc. **37**, 84–97 (1941)
16. Tanaka, Y.: Rubber Chemistry and Technology. **74**, 355–375 (2001)
17. Rublon, P., et al.: Eng. Fract. Mech. **123**, 59–69 (2014)
18. Göritz, D., Müller, F.H.: Kolloid-Zeitschrift und Zeitschriftfür Polymere. **241**, 1075–1079 (1970)
19. Le Cam, J.-B., Albouy, P.-A., Charlès, S.: Review of Scientific Instruments. **91**, 044902 (2020)

Chapter 7
Evaluating Pulse Simulator Using Fluorescent DIC

Chi-Hung Hwang, Rui-Cian Weng, Yen-Pei Lu, Wei-Chung Wang, Tzu-Yu Kuo, and Chun-Wei Lai

Abstract The artery vessel simulator mimics the pulse using a membrane pump to drive water flow through the PDMS phantom intermittently to simulate the blood flows in the artery vessel. The DIC method performs the evaluation, where the images used for DIC calculation are captured by a single-camera stereo-imaging system, which can eliminate the time synchronization issue. The fluorescent medicine FITC is used to generate the random patterns on the surface of a PDMS phantom, where the PDMS phantom is part of the artery vessel simulator. The DIC method determines the displacement and strain field introduced by the interaction between water and the PDMS phantom to know the mechanical behavior of the simulator. In this study, two different imaging modes are implemented to see the time various mechanical behavior of water-PDMS phantom interaction and the simulator's stability. Some challenges for using FITC to the PDMS would be highlighted at the end of this paper.

Keywords Artery vessel · Simulator · Fluorescence · Digital image correlation

7.1 Introduction

The digital image correlation method (DIC) is used for determining the object deformation or movement by tracking random patterns among images. The random marks either generated artificially or native posed are the must for performing the DIC method. While utilizing the DIC for various applications, the contact force is inevitably generated while rendering artificial random marks on the object surface. Thanks to some adequate methods, such as spraying, the force introduced while generating patterns are considered tiny and ignorable and always regarded as non-contact measurement methods. Nowadays, the DIC is widely used for different objects, such as determining the mechanical behaviors for common mechanical materials, constructing materials, soft materials, biomaterials, and others. One of the DIC practical applications is to determine the mechanical properties of biomaterials. During the last two decades, a tremendous amount of research on applying DICs to biomaterials has been reported, such as determining the mechanical properties of cells by tracking fluorescent particles [1], evaluating the mechanical behaviors of muscles [2–4], and evaluating the ventricular volume change after cardiac surgery [5–7]. This study is part of a project on developing a DIC system for general surgery to detect the possible vessel leakage and the mechanical behaviors of the organ after suture to replace the existing methods such as using radiopharmaceuticals and radiation detectors for vessel leakage detection.

While applying DIC to the in vivo applications, the compound used for generating random marks must be biological toxicity-free, and the system must be adequately evaluated to prove that the system performance meets the requirement, and the system configuration is well developed before either animal tests or clinical trials. Thus, to assess the imaging system design of a DIC system for the possible surgery applications, a simple simulator to simulate the mechanical interaction between the vessel and the tissue above the artery is essential. Therefore, the purpose of this study is to evaluate the artery simulator using DIC stereo-DIC method with the random surface pattern generated by fluorescent medicine before finalizing the system design.

C.-H. Hwang (✉) · R.-C. Weng · Y.-P. Lu
National Applied Research Laboratories, Taiwan Instrument Research Institute, Taipei, Taiwan
e-mail: chhwang@narlabs.org.tw

W.-C. Wang · T.-Y. Kuo · C.-W. Lai
Department of Power Mechanical Engineering, National Tsing Hua University, Hsinchu, Taiwan

© The Society for Experimental Mechanics, Inc. 2022
S. L. B. Kramer et al. (eds.), *Thermomechanics & Infrared Imaging, Inverse Problem Methodologies, Mechanics of Additive & Advanced Manufactured Materials, and Advancements in Optical Methods & Digital Image Correlation, Volume 4*, Conference Proceedings of the Society for Experimental Mechanics Series, https://doi.org/10.1007/978-3-030-86745-4_7

7.2   The Experimental Setup and the Artery Simulator [8]

Figure 7.1 shows the experimental setup used in this study, consisting of a single camera stereo-DIC imaging system and an artery simulator. In this study, FITC (fluorescein isothiocyanate) is adopted to generate the artificial random pattern on the object surface. FTIC can be excited at a peak wavelength around 495 nm and emits light at the peak wavelength of 519 nm. A self-made LED light source with the center wavelength around 475 nm is implemented to excite the FITC, and a long-pass filter with a cut-on frequency at 500 nm is placed in front of the lens for blocking the excitation light. A signal camera stereo-imaging optical imaging system is implemented to be the DIC imaging system that can avoid the synchronization problem. Regarding the artery simulator, as reported previously [8], consists of a membrane pump, a PDMS phantom, and a computer for handling ECG signal generation and ECG signal detection. The membrane pump is used to draw water out of a tank and then the water is forced to go through the PDMS phantom with water pressure varied with the time; afterwards, the water will be sent back to the tank for completing the water circulating. A PDMS-made phantom is casted in a hollow cylindrical shape, whose height is 100 mm with ∅ 50 mm and ∅ 70 mm inner and outer diameter, a 4 mm diameter with 3 mm depth below the outer surface is cast to simulate a medium-size near-surface artery vessel inside an organ.

A computer-generated ECG signal drives the pump to allow the water pressure to vary with time. Then the water will go through the PDMS phantom intermittently to simulate the artery behaviors. Meanwhile, the generated ECG signal will also be injected back to the computer to detect the R-peaks and T-peaks for sending trigger singles to the camera for taking images. In the experiment, the computer-simulated ECG signal is first sent out to a DAQ; DAQ sends back the ECG signal to the computer to find the ECG R- and T-peaks. Then the trigging signal is sent out to active the camera. That means, the single-camera stereo-DIC imaging system is used to capture images after the computer detects the ECG R-peaks and/or T-peaks. The time-lapse for imaging after the detection of R-peak/T-peak has been evaluated to be 15 ms [8].

In this study, the first taken R-peak image is always used as the reference image. Then two imaging operation modes are implemented to capture the operation images; the first mode is catching T-peak images after the first R-peak is detected; the second mode is continuously imaging immediately after the reference image is captured, as shown in Fig. 7.2. Regarding the time interval for continuously imaging is set to be 20 ms and the shutter time is 3 ms. Simultaneously, the membrane pump is derived by a home-made driver with the computer-generated ECG signal. The membrane pump is then intermittently drawing water out of a tank and then driving the water flow through the connecting vessels, the PDMS artery vessel phantom. It then flows back to the water tank to complete the water circulation through the system. The maximum volume of water the membrane pump can circulate is 400 ml/min. Considering the experiment is a dynamic case. Therefore, the exposure time is set to be a short one that is not good fluorescent practice and makes the measurement a great challenge. In this evaluation study, the VIC-3D of Correlated Solutions is used for displacement and strain calculation. The displacement and strain fields are described according to the coordinate system defined on the PDMS phantom, as shown in Fig. 7.3. The x-direction is normal to the direction of water flowing, and the y-direction is parallel to the inner vessel with the positive direction aligning to the water out. Meanwhile, in this study, the PDMS phantom is free placed with no strong fasten applied.

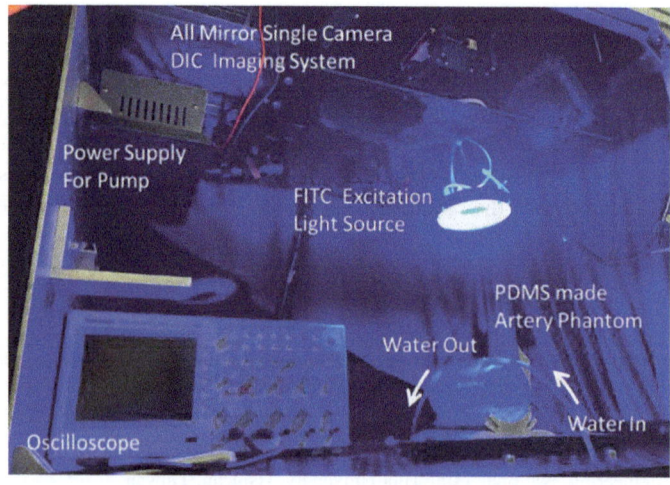

Fig. 7.1 Experimental Setup

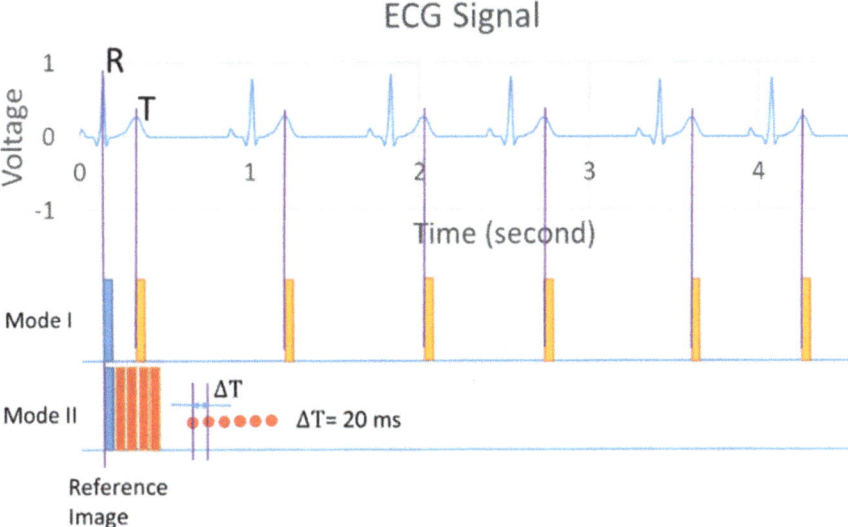

Fig. 7.2 Two imaging modes in this study

Fig. 7.3 PDMS phantom
and the coordinate system

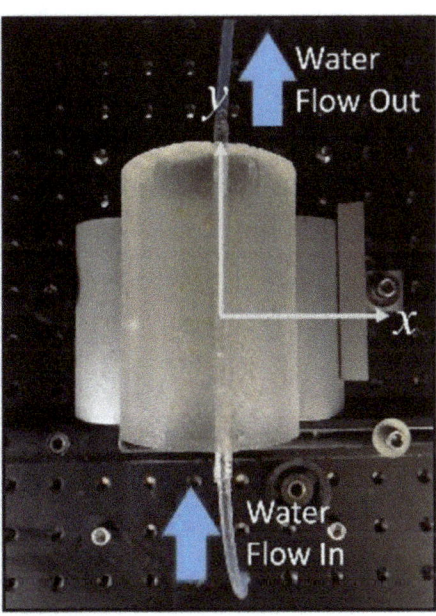

7.3   Results and Discussions

In this study, the FITC is first sprayed to the surface of the PDMS phantom; different from the typical applications of FITC on cells or tissues, the FITC emission light attenuates far more quick than the biomaterials. The random surface pattern does not last as long as the traditional paint spraying method; therefore, images taken by two image modes will have different surface artificial random patterns. The discussion on the displacement and strain field determined by two different imaging modes will not be quantitative but qualitative.

Figure 7.4 presents five selected displacement fields of the PDMS phantom as the images are continuously recorded after the first computer-generated ECG R-peak detected. From the averaged U-displacement plots, shown in Fig. 7.4a; the pumping water introduces the phantom to vibrate, and time-sequentially causes five specific displacement packages to the PDMS phantom, as shown inside the green-box of Fig. 7.4a, during the time the imaging system takes the first 180 images. Similar results can also be found at V- and W-averaged displacement plots; however, the V-displacement plot doesn't show the displacement packages as clear as the U- and W-averaged displacement plots. For U-displacement shown in Fig. 7.4a, although the averaged-displacements are different, the image frames #10 and # 25 have similar patterns. Those two

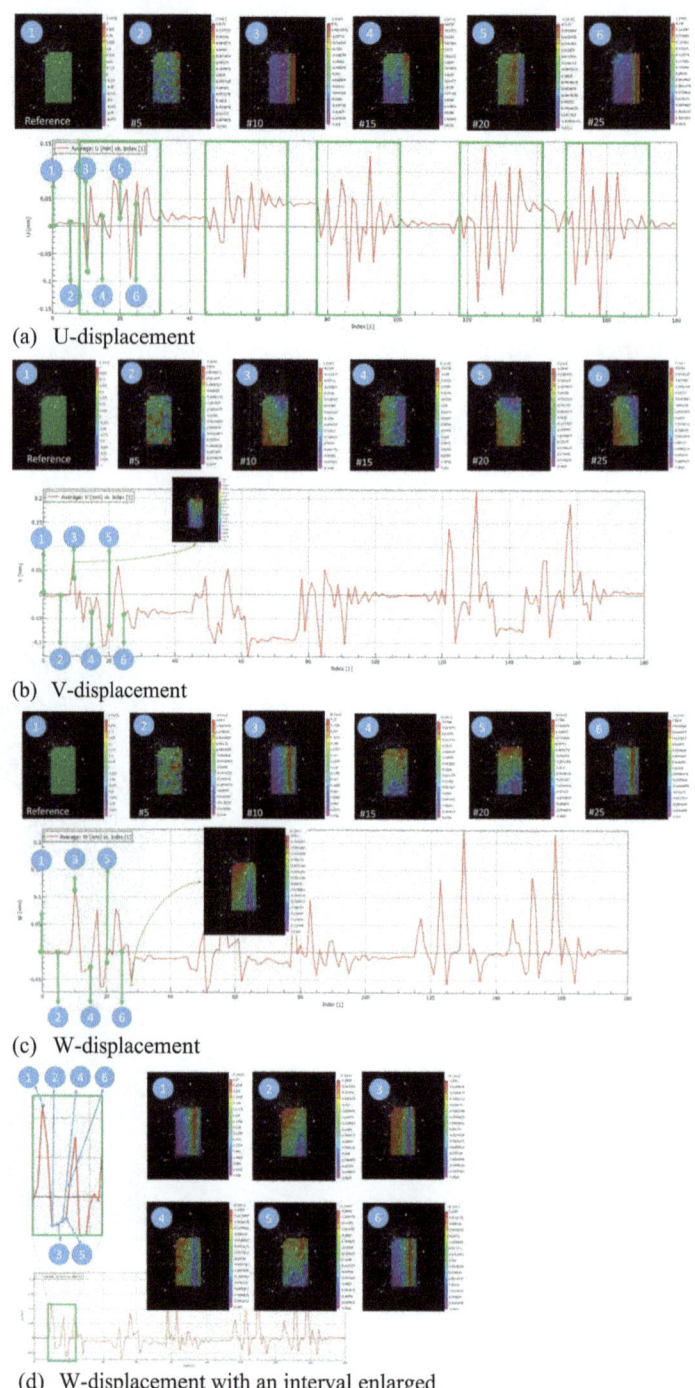

(a) U-displacement

(b) V-displacement

(c) W-displacement

(d) W-displacement with an interval enlarged

Fig. 7.4 The displacement field of the PDMS phantom subjected to water flow intermittently pumping through (**a**) U-displacement, (**b**) V-displacement, (**c**) W-displacement, (**d**) W-displacement with an interval enlarged

displacement fields show the right-hand side of the PDMS phantom inner vessel tends to move to the right and on the inner-vessel left-hand side tends to move to the left or less to the right; that means the surface of the inner vessel tends to expend. As for the V-displacement field shown in Fig. 7.4b, since the direction is parallel to the flow direction of the pumped water, without significant displacement, the results seem reasonable. However, while taking a close look at the top part of the PDMS, some remarkable local displacement can be observed, for example, images of frames #10, #20, and #25. Considering the region is of the same as the exit of the vessel to the connecting pipe. Therefore, the displacement would be regarded as introduced by the drag force. Regarding the W-displacement shown in Fig. 7.4c, image frames #10 and # 25 reveal that the PDMS surface above the vessel tends to move upper; combining this observation with the U-displacement fields discussed previously, it could easily be concluded that the water at these moments are pumping into the vessel. Besides, frames #15 and #20 also present local displacement around the inlet of the vessel. To understand this phenomenon, some of images between frames #15 and #20 are inspected with details; as shown in Fig. 7.4d, the interaction between intermittently pumping water and the PDMS phantom can be clearly observed from the series six frames. The water flowing into the vessel with relatively high pressure will cause the PDMS surface above the vessel to expand; then the water flows out and pumping pressure drops, then a local low-pressure region around the inlet of the vessel will build and extend to the exiting side of the vessel as the water flows out; after then, the volume of the vessel is no more fully filled with water; while the pump is active again, the water will be accelerated by pumping force and flow into the vessel to fill the space, then the inner pressure of the vessel would be increased rapidly.

Regarding the strain fields e_{yy} and e_{xy}, they do not present apparent patterns for the first 25 frames, as shown in Fig. 7.5b, c. However, e_{xy} shows a very interesting local strain at the middle of the vessel, but no proper physical model can adequately

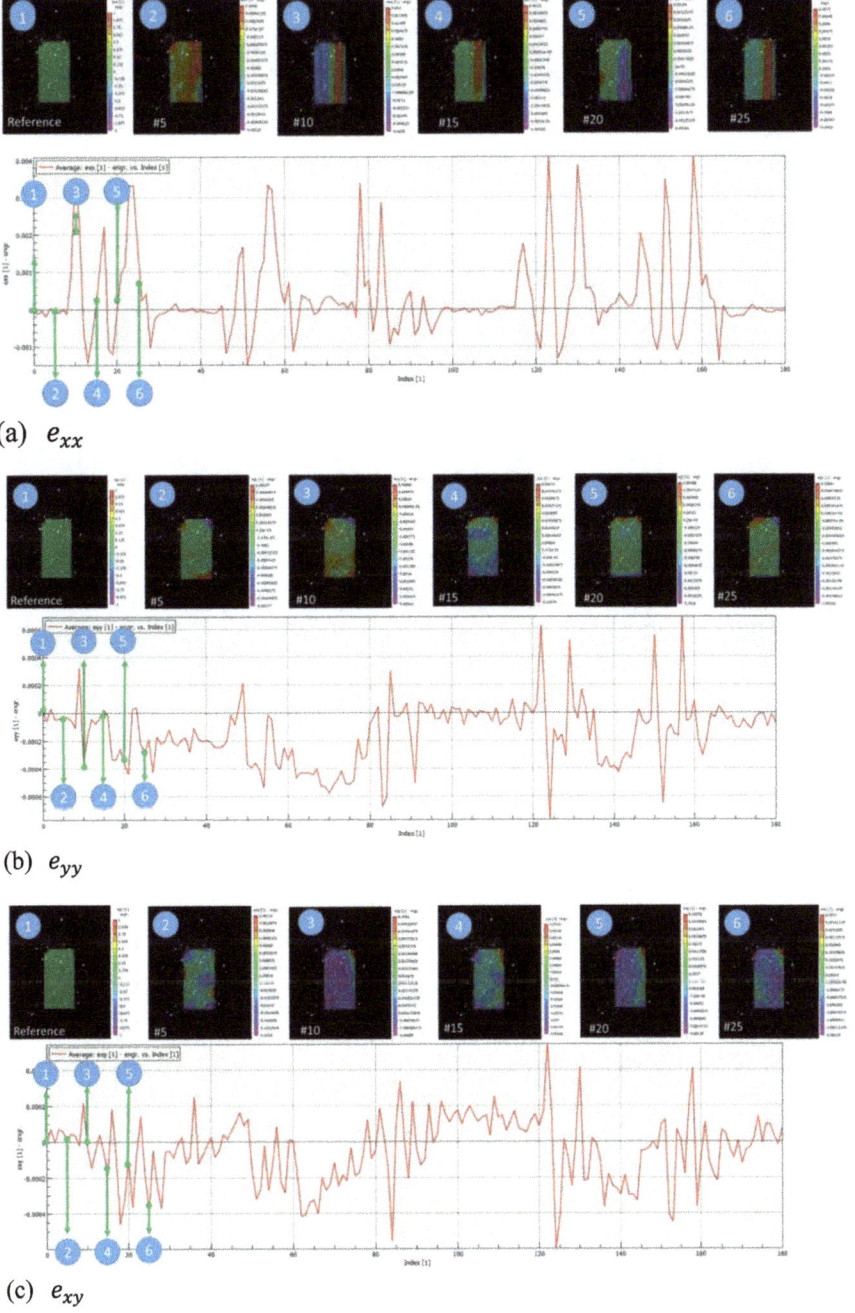

(a) e_{xx}

(b) e_{yy}

(c) e_{xy}

Fig. 7.5 The strain field of the PDMS phantom subjected to water flow intermittently pumping through (**a**) e_{xx},(**b**) e_{yy},(**c**) e_{xy}

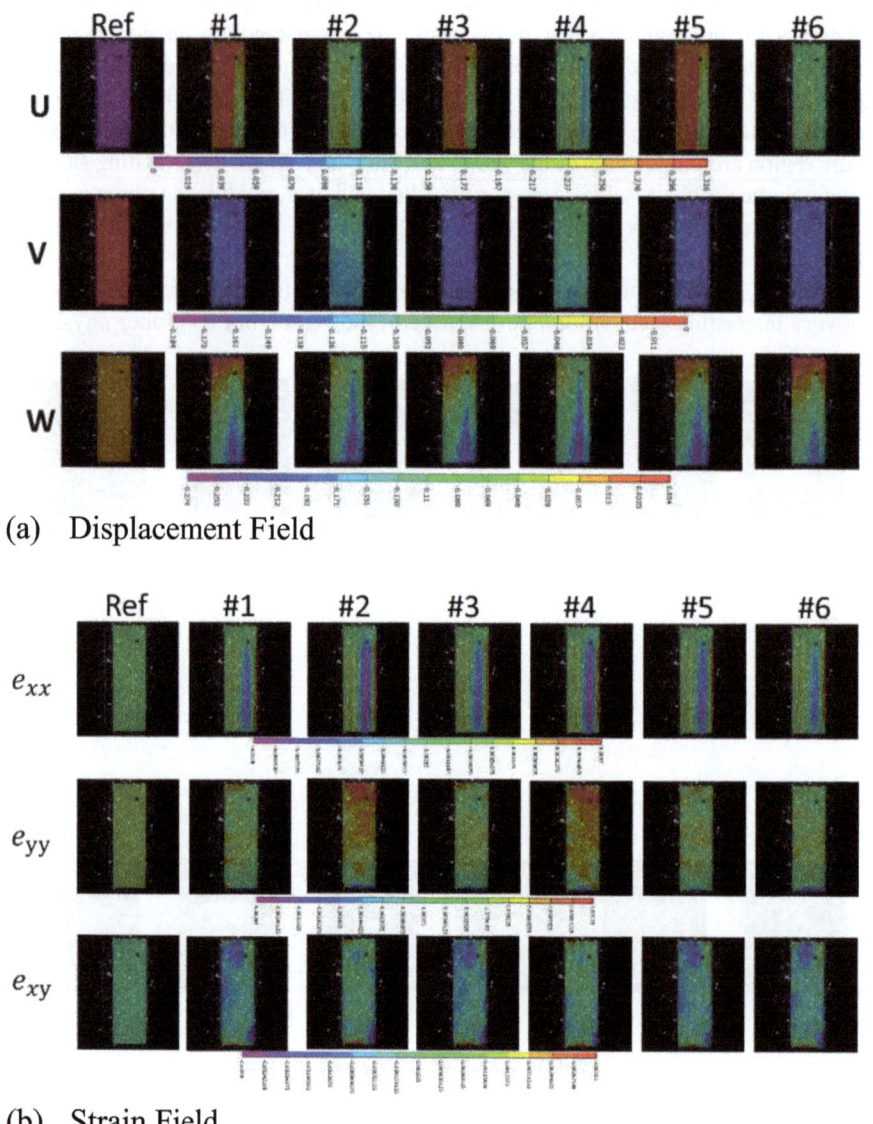

(a) Displacement Field

(b) Strain Field

Fig. 7.6 The displacement and strain field of the PDMS phantom subjected to water flow intermittently pumping through (**a**) displacement field, (**b**) strain field

explain this strain field. Based on the displacement field discussed previously, the U-displacement fields shown in frame #10 and #25 indicate the surface of PDMS above the vessel tends to expend; intuitively, the strain fields e_{xx} of those two frames are expansive strain as shown in Fig. 7.5a. Aside from the previous results, frame #20 of Fig. 7.5a shows a compressive strain field above the vessel that is not observed from the corresponding U-displacement field shown in Fig. 7.4a; obviously, the strain field e_{xx} provides more data, sensitive, than the U-displacement field.

The other imaging mode is taking the first image as reference one while the ECG R-peak is detected. Then the imaging system continuously takes images as ECG T-peaks caught. The fetched images are then used for displacement and strain calculation by the DIC method. The calculated strain and displacement fields are presented in Fig. 7.6a and Fig. 7.6b, respectively. Different from the previous results, the displacement and strain fields here are repeatedly calculated as T-peaks detected. Therefore, if the ECG signal is correctly repeated and the output power of the pump is well controlled, then the resulting displacement and strain would be almost the same; that means this image mode can be used to evaluate how stable the simulator is. As shown in Fig. 7.6, the displacement and strain fields can be divided into two groups. For the first group, the displacement and strain fields of #1, #3, and #5 T-peaks are almost identical, as for the other group is the displacement and strain field determined using the images taken as #2 and #5 T-peaks detected. Even though the displacement and strain fields seem can be divided into two groups, the displacement and strain magnitudes are not varied tremendously. Based on the resulted displacement and strain field, a small output performance variation existing in the simulator can be evaluated. Still, the influence on the mechanical interactions between the PDMS and the water introduced by the variation of the simulator is limited.

Regarding the displacement field conducted by the PDMS-water interaction, the U-displacement indicates the PDMS tends to move to the right. The region above the vessel tends to shrink because the U-displacement on the right-hand side above the vessel is smaller than the displacement magnitude of the left-hand side. As for the V-displacement, the plots show that the PDMS would be uniformly moved towards the connecting pipe on the inlet side; however, the physical model for explaining the displacement is not available now. From the W-displacement field, the region above the vessel tends to move down at the inlet side and tends to move up at the exit side; this indicates pressure inside the vessel dropped near the inlet side and the whole PDMS is left up at the exit side. The displacement indicates pressure inside the vessel dropped at the inlet side and then causes the surface above the vessel move downwards. As for the exit end, the displacement might be caused by superposing a local expansion and a vibration. Regarding the local expansion, since the diameter of connecting pipe is smaller than the through hole of PDMS (to simulate the vessel), the inner pressure of the vessel is locally high and might cause the region around the exit side to expand.

The strain fields are present in Fig. 7.6b, the e_{xx} of all six successive frames show the normal strain above the PDMS vessel are all in compressive, and the magnitudes are highest at frames #2 and #4. Regarding the strain field e_{yy}, frames #1, #3, #5, and #6 show that most of the PDMS phantom is subjected to small compressive strain but relatively small extensive strain in the other regions. The PDMS upper part shows a relatively high extension strain value for e_{yy} of the #2 and #4 frames. Besides, while inspecting Fig. 7.6b with care, the strain field around the inlet and exit edges shows a small compressive and extensive e_{yy}, respectively; the strain e_{yy} around the edge might be caused by the pressure drop at the inlet, and the high pressure introduced by diameter mismatch at the outlet side. Regarding the shear strain e_{xy}, frames #1, #3, # 5, and #6 indicate the upper-left, and bottom-right of the vessel are subjected to the shear strain locally but different sign. Similar results can be observed from frames #2 and #4; however, the shear strains at the upper-left part are relatively small compared with the other cases. More interest, when the e_{xy} is relatively uniform, the e_{xx} above the vessel apparent more strong compressive strain.

From the above discussion, the developed artery vessel simulator can provide relatively stable mechanical behaviors at each T-peak from the above discussion. The time-sequential mechanical behaviors are varied for the V-displacement, e_{yy}, and e_{xy}.

In this study, the developed artery-vessel simulator is deeply investigated using a self-developed single-camera stereo-DIC system. The random surface pattern of the artery-vessel simulator is prepared using FITC. FITC is a fluorescent substance that has been widely used for cell and tissue studies. Typically the FITC is prepared by dissolving the powder with water. In this study, the FITC is sprayed on the surface of PDMS to provide a random pattern for tracking; however, the water will evaporate rapidly if the size of the drop is small. And then the emitting light will attenuate. In contrast, the drop will move due to the self-weight of the drop is large; the drop will also move while subjected to the force as the water intermittently flow through the PDMS that will introduce the pattern changed. Therefore, different from the traditional artificial random pattern, the FITC random pattern cannot last as long as the random pattern generated with ink and paint. Although many fluorescent medicines can be used to generate random patterns for DIC use, more works are required for handling the problem mentioned.

7.4 Conclusions

In this study, FITC is used to generate the random pattern on the PDMS surface. Then the random patterns are tracked from frame to frame using the DIC algorithm for evaluating the associated displacement and strain among frames for evaluating the mechanical behaviors of the proposed artery-vessel simulator. The proposed simulator has proved to be relatively stable for simulating the mechanical behaviors among the first detected ECG R-peak and the following T-peaks. As for time-sequential mechanical behaviors, the simulator might need additional adjustment to make the mechanical behaviors more consistent. Although the FITC can be used for DIC application, the emitting light attenuation because of the water evaporation needs to be solved.

Acknowledgments This paper was supported in part by the Ministry of Science and Technology, Taiwan (Grand no. MOST-107-2221-E-492-012 and MOST-108-2221-E-492-022-MY2).

References

1. Berfield, T.A., et al.: Fluorescent image correlation for nanoscale deformation measurements. Small. **2**(5), 631–635 (2006)
2. Praveen, G.B., Raghavendra, S., Chang, V.I.C.: An analysis of leg muscle stretch using 3D digital image correlation. Int J Organiz Collective Intell. **7**(3), 30–43 (2017)
3. Xue, Y., et al.: High-accuracy and real-time 3D positioning, tracking system for medical imaging applications based on 3D digital image correlation. Opt. Lasers Eng. **88**, 82–90 (2017)
4. Solav, D., et al.: MultiDIC: an open-source toolbox for multi-view 3D digital image correlation. IEEE Access. **6**, 30520–30535 (2018)
5. Hokka, M., et al.: In-vivo deformation measurements of the human heart by 3D digital image correlation. J. Biomech. **48**(10), 2217–2220 (2015)
6. Hokka, M., et al.: DIC Measurements of the human heart during cardiopulmonary bypass surgery. In: Mechanics of Biological Systems and Materials, pp. 51–59. Springer, Cham (2016)
7. Soltani, A., et al.: An optical method for the in-vivo characterization of the biomechanical response of the right ventricle. Sci. Rep. **8**(1), 1–11 (2018)
8. Hwang, C.-H., et al.: Evaluating the application of DIC on heartbeat detection by using a self-developed artery vessel simulator, Mechanics of Biological Systems & Micro-and Nanomechanics, Volume 5 of the Proceedings of the 2020 SEM Annual Conference & Exposition on Experimental and Applied Mechanics (2021)

Chapter 8
Predicting Temperature Field in Powder-Bed Fusion (PBF) Additive Manufacturing Process Using Radial Basis Neural Network (RBNN)

Ehsan Malekipour, Homero Valladares, Suchana Jahan, Yung Shin, and Hazim El-Mounayri

Abstract Avoiding or eliminating thermal abnormalities in powder bed fusion (PBF) is critical since the abnormalities can lead to a higher failure rate of printing complex parts, a longer manufacturing lead time, and/or additional post-processing. Controlling the thermal evolution of the process can hinder or minimize some of the most frequently encountered thermal abnormalities. To achieve such an objective, the prediction and control of temperature distribution throughout an exposure layer is a crucial step. The generation of uniform temperature distribution throughout the printed layers and the avoidance of overheated zones are two primary sub-objectives for controlling the thermal evolution of the process. However, the complex and non-linear nature of the process has limited the ability to derive a universal analytical equation to correlate the process parameters with the thermal distribution of a printed layer. Laser specifications such as laser power and scanning speed are among the main process parameters that predominantly govern the temperature distribution throughout the layer. In this paper, we employ an artificial neural network (ANN) to correlate laser power with the temperature of the printed area around the melt pool in Inconel 718. In our first variant, we investigate the effectiveness of using the multilayer perceptron Radial Basis Neural Network (RBNN) to model the function for predicting the temperature distribution for various laser power. We use the Rosenthal equation to generate adequate inputs-outputs for training our function. We then compare the output with the simulation results for five different laser powers. The results show that the function was trained successfully with a low mean square root error of 9.7157 using 2000 samples, a wider gap exists between the trained function and the simulated data. In the second variant, we use a recurrent neural network (RNN), which enables temporal histories to be used for training. To fulfill such objective, we acquire real thermal data using a photon-counting IR camera for different printed layers. This step allows the training of a function to predict the temperature distribution precisely for different laser power and thermal history. As future work, we will employ the function to adjust the laser power to minimize the overheated zones and distribute the temperature uniformly throughout each exposure layer.

Keywords Powder bed fusion process · Artificial neural network · FBFN neural network · Temperature prediction · Adjusting laser power

8.1 Introduction and Problem Definition

In recent years, the versatile metal additive manufacturing (AM) processes have gained significant interest and applications in various industries. As a promising technology offering endless opportunities to design and manufacture complex customizable parts that are almost impossible with traditional manufacturing, industries are adopting this technology and taking advantage of the digital product development, generative design, economic efficiency, and tool-free benefits offered by this technology. Integrating AM in the industry 4.0 workflow provides further significance and application for this technology. Among the different metal additive processes, the powder bed fusion (PBF) process is extensively studied and deployed as it offers advantages for manufacturing near fully dense parts with a wide range of material selection capability. In the past decade, aerospace, defense, and medical industries comprehensively used the PBF process to develop parts with high thermo-mechanical properties, porous structures, and low weight generative designs. Inconel 718 is complex precipitation

E. Malekipour · H. Valladares · S. Jahan (✉) · Y. Shin
Purdue University, West Lafayette, USA
e-mail: sjahan@iupui.edu

H. El-Mounayri
Purdue School of Engineering and Technology, Indianapolis, USA

© The Society for Experimental Mechanics, Inc. 2022
S. L. B. Kramer et al. (eds.), *Thermomechanics & Infrared Imaging, Inverse Problem Methodologies, Mechanics of Additive & Advanced Manufactured Materials, and Advancements in Optical Methods & Digital Image Correlation, Volume 4*, Conference Proceedings of the Society for Experimental Mechanics Series, https://doi.org/10.1007/978-3-030-86745-4_8

strengthening alloy that is known for its high temperature creep resistance, high cooling rates, and high strength properties [1, 2]. Due to typical high strength, fatigue, high temperature, and high-performance applications in the aerospace industry, Inconel 718 is a popular superalloy material choice [1].

Despite the widespread application, significant research, and incredible accomplishments in the PBF process, complete process, and quality control has yet to materialize due to the complex nature of the process. More than 50 parameters influence the thermal evolution, melt-pool dynamics, and the quality of the end-product. Among all the factors in the PBF process, laser power, scan speed, layer thickness, beam diameter, hatch spacing, and scan pattern are the crucial parameters with substantial impact on the thermal evolution, melt-pool dynamics, quality and process control. Controlling these parameters subsequently offers the control over the in-process defects, i.e., balling, lack of fusion, keyhole and crack formation, and geometric anomalies. Thermal history plays a crucial role in the defect generation and significantly affects the quality properties. Being a layer-based technology, monitoring, and controlling the thermal evolution history including heat-affected zones (HAZ) is crucial to obtain complete control over the defect generation and the overall process.

Lately, researchers have concentrated on studying the effects of thermal evolution on the defect generation and are working to monitor the thermal history of each layer, thus controlling the process parameters for the subsequent layers. Promoppatum et al. studied the thermal behavior of Inconel 718 by comparing the analytical Rosenthal solution and an FE model [3]. This proved that both analytical and FE model have close predictions but the thermal history predictions using analytical method is less accurate at high energy inputs as the analytical solution does not take heat losses into account leading to an assumption of low cooling rates. Many research studies used FE models to study the correlation between process parameters, thermal evolution, and defect generation ultimately providing insight into the necessary adjustments of the pre-process parameters to reduce the defects and to obtain uniform microstructure. The major drawback in the literature using solely FE models to study thermal evolution and defect generation is that these studies are incapable of capturing the in-process thermal history, layer-wise monitoring, and in-process control.

To overcome this, researchers directed their motivation to the data-driven techniques which offer real-time machine learning (ML) and neural network (NN) approaches for prediction of the thermal history of each layer in real-time. ML and NN approaches have promising application capabilities in capturing a large amount of data in real-time and in identifying the non-linear relation between process, material, and quality properties. As a viable option, ML and NN approaches are coupled with FE models and real-time thermal imaging, improving the prediction capabilities, eventually to achieve the complete control on the process parameters and defect generation. Many ML and NN architectures were developed in recent years leading the way for better non-linear relation derivability and data learning. Ohyung et al. and Christian et al. used convolutional neural networks (CNN), melt-pool images, and real-time captured data using a high-speed camera from the PBF monitoring system to train a function in predicting laser power and dimensional accuracy values [4, 5]. Using this model, the formation of defects can be identified on every specific layer and helps in adjusting the laser power and other process parameters. Ivanna et al. proposed a conceptual framework, combining machine learning, statistical analysis, and mathematical modeling to control the quality of the parts produced [6]. This framework highlights the importance and benefits of using ML techniques together with mathematical modeling. Also, the two common aspects in the studies using ML are distinctly stated, i.e., the motivation is routed because of the lack of knowledge and understanding of the effect of multiple parameters on the part quality, and the generalization challenges of the ML algorithm techniques because of the small sets of data and high cost for obtaining experimental data. Arindam et al. proposed a framework for developing a real-time temperature prediction using ML techniques and FE model for data generation and achieved a mean absolute error of 1% [7]. Generalized Analysis for Multiscale Multi-Physics Application (GAMMA), an FE model-based time-dependent heat equation method, was used to generate the datasets required for training the ML algorithm. This work only used the partial differential equation-based data generation and ML without any experimental comparison leading to the uncertainty of the results published.

Seulbi et al. used the ML models to predict the melt-pool geometry and successfully demonstrated that modern data analytical approaches will accelerate the melt-pool control and process parameters optimization [8]. In this work, they conducted single-track experiments on Alloy 625 and Alloy 718 powders and generated the data for training the ML algorithms. Using correlation analysis, the ML algorithm models were optimized and improved for better predictions. Similarly, Alessandra et al. and Bodi et al. demonstrated that the deep CNN and ML models can characterize the layer-wise images and can effectively identify the in-process defects generated due to abnormalities in the process [9, 10]. The developed deep CNN model was able to capture the data of each layer and fuse the layers patterns to classify the defects and progression through layers. Xinbo et al. presented the current application, integration, progress, and future trends of different ML and NN techniques in attaining real-time control of the process [11]. They outlined the current challenges in the interdisciplinary area of using ML and NN techniques and how new models are being developed continuously to address the existing challenges. The crucial challenges facing the adoption of ML techniques are restrictive small datasets, data labeling errors, knowledge gap in selecting good features for ML or NN techniques, and the problematic overfitting and underfitting. With the smaller dataset, the margin of error in predictions is high leading to uncertainty in the model predictions.

This can be avoided by either having more experimental data or by using generative models to enlarge the dataset synthetically. Overfitting and underfitting are the major challenges in attaining the stability of the function and error rate leading to the generalization of the trained model [11]. To address all these issues, scholars are employing different ML and NN techniques in their work to recognize which ML/NN technique has the best possibility in dealing with the challenges and attaining the generalization of the developed model irrespective of the process–parameter–quality relationships.

In this work, a radial basis neural network (RBNN) is developed to correlate laser power with the temperature of the printed area around the melt pool in Inconel 718. Firstly, a function is modeled and trained using RBNN for predicting the temperature distributions for various laser power. The inputs-outputs for training the function are generated using a standardized Rosenthal equation. This equation is extensively used in the fusion welding process and is currently being used for Metal AM processes due to the close similarities between both processes. This equation considers the thermal properties, scan speed, and heat source distance and predicts the thermal characteristics in the process. The trained model is compared with the simulation results with five different laser powers. The main objective of this research is to initially train the model for predicting the thermal characteristics using laser power and vice versa from the data generated using the Rosenthal equation. Using this trained model, the future work aims to acquire real-time data and optimize the function for real-time application. Ultimately, this research will develop a model that will be employed in real-time monitoring of overheated zones simultaneously adjusting the laser power improving the uniformity of the temperature distribution throughout each exposure layer.

8.2   Methodology

Laser specifications in powder bed fusion process such as laser power and scanning speed are among the major process parameters that predominantly govern the temperature distribution throughout the laser. In this paper, the Rosenthal equation is employed to calculate the temperatures for different laser parameters. Rosenthal originally developed the analytical method to predict the temperature history in fusion welding. However, due to the similarities between fusion welding and Selective Laser Melting (SLM), the Rosenthal equation is employed to predict the thermal characteristics in SLM process here. The Rosenthal equation is as follows:

$$T = T_o + \frac{\lambda P}{2\pi k r} \, exp \left[-\frac{V(r + \xi)}{2\alpha} \right]$$

Here T_0 = temperature at locations far from the top surface, k = thermal conductivity, V = scanning velocity, α = thermal diffusivity, ξ = moving co-ordinate of (x-Vt), r = distance from the heat source.

The Rosenthal equation was derived based on the following assumptions:

- The thermophysical properties, including thermal conductivity, density, and specific heat, are temperature independent.
- The latent heat due to phase change is not included.
- The heat source is a point source.
- Heat losses from surface convection and radiation are not considered, and convection in the liquid pool is neglected. Therefore, the heat transfer is governed purely by conduction.

However, we can use this equation to correlate T and P, but it is not completely accurate. Yet, as we are just trying to use this equation to generate data for training ANN function and evaluate our results, this method is an easier and "cheap" way for data collection/generation.

By relating laser power (P) and distance® from the thermal flux, the generated data is plotted in the following Fig. 8.1. The range for P and r as follows:

100 Watt $\leq$ P $\leq$ 250 Watt
0.1 mm $\leq$ r $\leq$ 0.2 mm.

Using our code, we tried to make the results as similar as Rosenthal Temperature Distribution Function. The ANN function parameters are adjusted as follows:

1. Spread.
2. Number of nodes.
3. Number of neurons.
4. Number of Training samples.

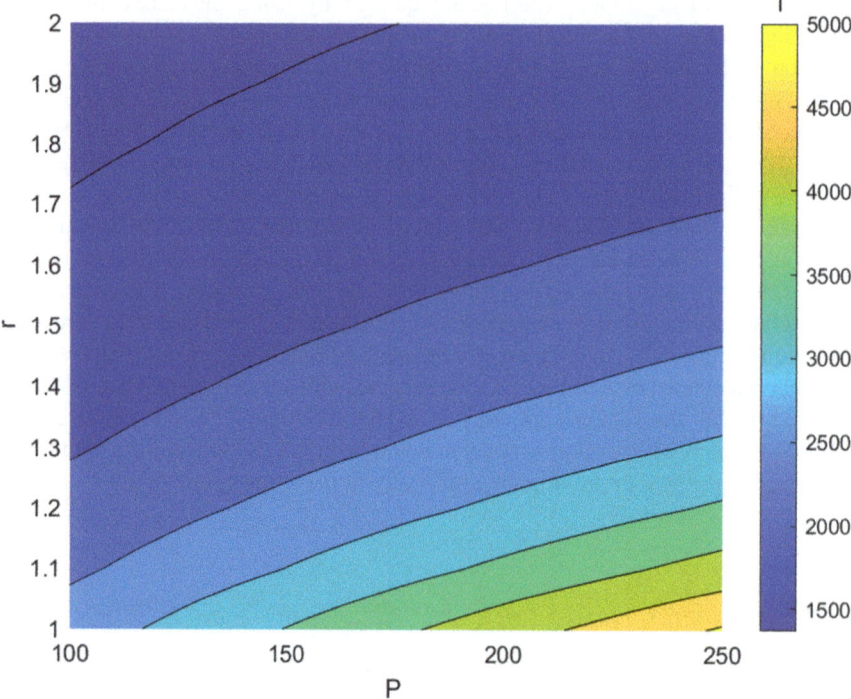

Fig. 8.1 Generated data using Rosenthal equation

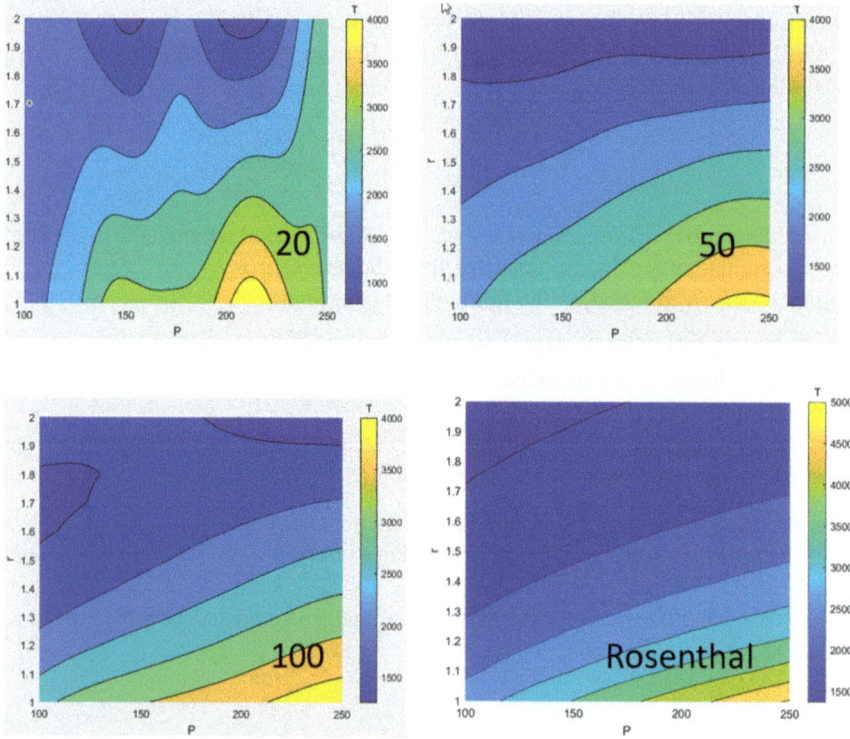

Fig. 8.2 Training of ANN and adjustment

The following plots are obtained that shows the adjusting the spread to obtain similar distribution as the Rosenthal equation (Fig. 8.2).

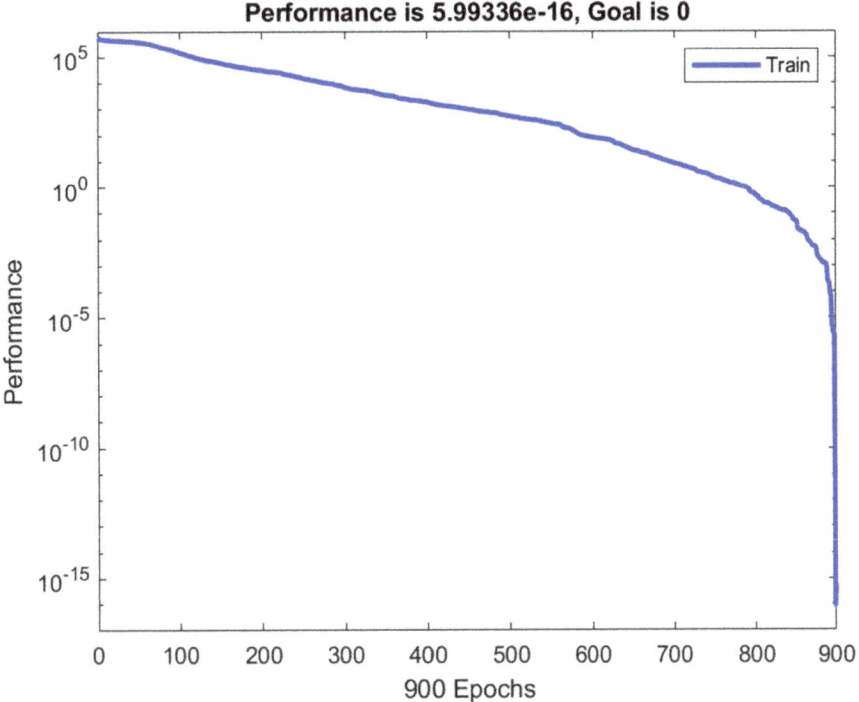

Fig. 8.3 Training test result for Test1

8.3 Results and Discussions

8.3.1 Training Results

For training purposes, two tests were conducted. In the first set, the number of training samples, n, is 900 and number of validation samples, m, is 300. On the other hand, the second test contains 2000 training samples and 600 validation samples. The goal is 0, i.e., to make the error as zero. We can see in second test the total number of samples is largely increased. The testing is done using convergence curves method. It compares the number of training samples and validation samples. In Figs. 8.3 and 8.4, the number of Epochs (iterations) is the number needed to find optimized parameters for artificial neural network (ANN). It is evident that different number of samples requires different number of epochs to converge.

8.3.2 Graphical Approximation Results

In this section we describe how the real values and predicted values compare in both tests. In the first test with $n = 900$ and $m = 300$, the comparison test is shown in Fig. 8.5. This is a comparison plot that shows relation or comparison between real and predicted values using ANN method. The more it aligns with 45° line, the more accurate the results are of the artificial neural network.

In the second test, when the number of samples are increased (here $n = 2000$, $m = 600$), we can achieve more accurate results compared to the first test. It is shown in Fig. 8.6.

In Fig. 8.7, the test results are shown as a comparison between ANN for different number of samples. The left figure for the first test is too noisy, which corresponds to test 1 ($n = 900$, $m = 300$). The right figure has lesser noise as it had higher sample number ($n = 2000$, $m = 600$). The Root mean square error for these two tests are 81.0886 and 9.7157, respectively, which is a significant improve. However, from this figure, we can state that when the number of samples is 1200, the prediction model is sufficient.

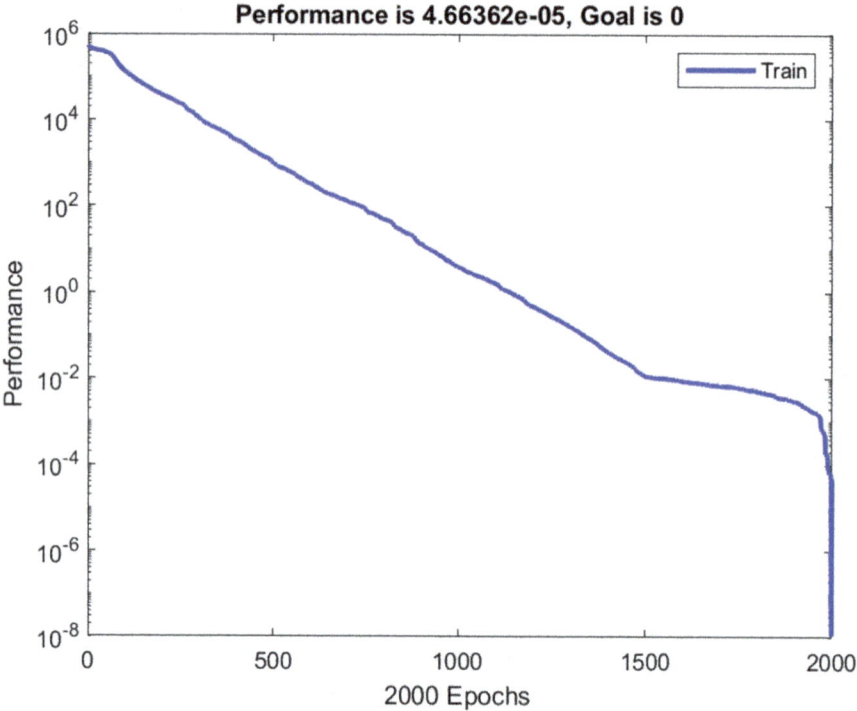

Fig. 8.4 Training test result for Test2

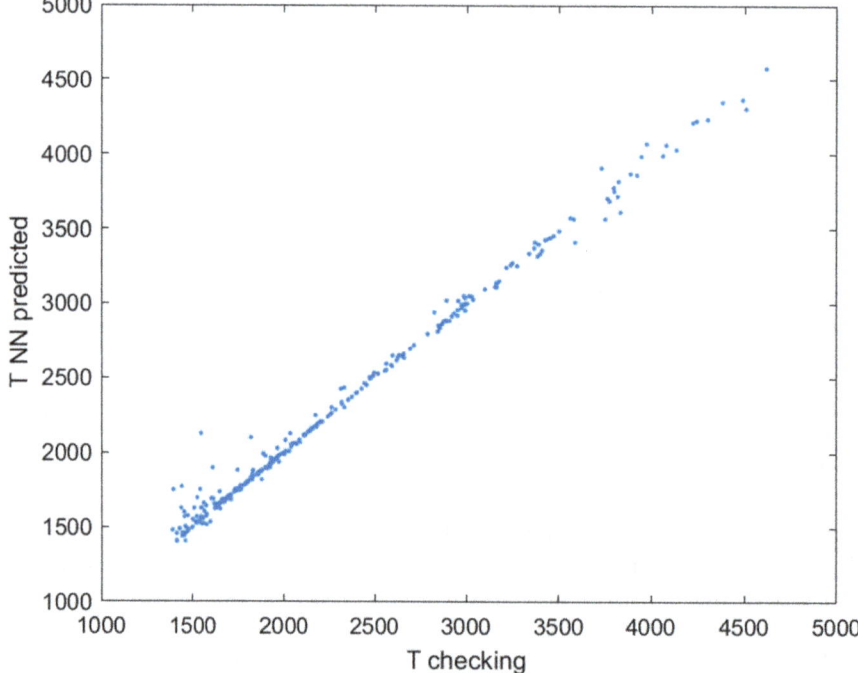

Fig. 8.5 Result comparison for Test 1

8.4 Conclusions and Future Work

We can conclude that more samples for training can be used to obtain more accurate results. However, it is significantly more time consuming for the training data. However, we also need to consider overfitting cases, if we increase the number of training samples. In this work, we have tested the optimized number of nodes and neurons with lower number of training

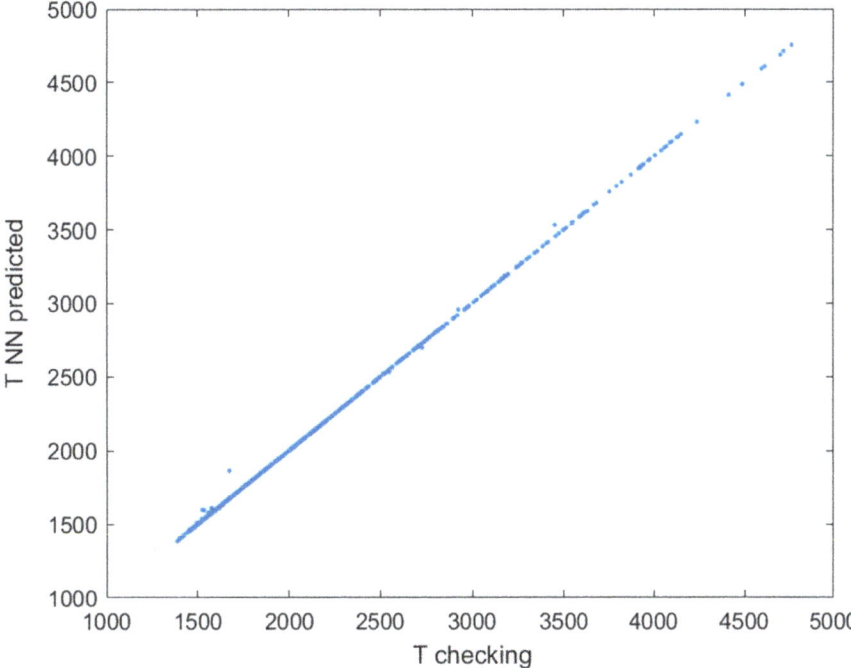

Fig. 8.6 Result comparison for Test 2

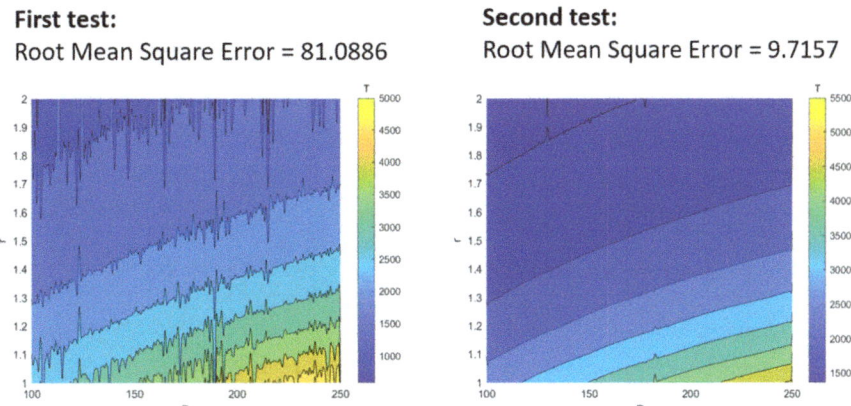

Fig. 8.7 Test results and comparison

samples, and then increased the number of samples later. Increasing the spread increases the precision of the function and the curve becomes much smoother. Again, changing the number of nodes after 64 does not affect the precision of results significantly. However, The Goal converges to zero and the mean square root error (MSRE) is very small with using 2000 samples and after 2000 Epochs. Moreover, there are few assumptions for derivation of the Rosenthal equation; hence we actually need experimental data to training a function that will virtually accord with reality.

References

1. Xu, Z., et al.: Creep behaviour of inconel 718 processed by laser powder bed fusion. J. Mater. Process. Technol. **256**, 13–24 (2018)
2. Barros, R., et al.: Laser powder bed fusion of inconel 718: residual stress analysis before and after heat treatment. Metals. **9**, 12 (2019)
3. Promoppatum, P., et al.: A comprehensive comparison of the analytical and numerical prediction of the thermal history and solidification microstructure of Inconel 718 products made by laser powder-bed fusion. Engineering. **3**(5), 685–694 (2017)
4. Kwon, O., et al.: A convolutional neural network for prediction of laser power using melt-Pool images in laser powder bed fusion. IEEE Access. **8**, 23255–23263 (2020)

5. Gobert, C., et al.: Application of supervised machine learning for defect detection during metallic powder bed fusion additive manufacturing using high resolution imaging. Addit. Manuf. **21**, 517–528 (2018)
6. Baturynska, I., Semeniuta, O., Martinsen, K.: Optimization of process parameters for powder bed fusion additive manufacturing by combination of machine learning and finite element method: a conceptual framework. Procedia CIRP. **67**, 227–232 (2018)
7. Paul, A., Mozaffar, M., Yang, Z., Liao, W. K., Choudhary, A., Cao, J., Agrawal, A.: A real-time iterative machine learning approach for temperature profile prediction in additive manufacturing processes, IEEE International Conference on Data Science and Advanced Analytics (DSAA) (2019), October. p. pp. 541–550
8. Lee, S., et al.: Data analytics approach for melt-pool geometries in metal additive manufacturing. Sci. Technol. Adv. Mater. **20**(1), 972–978 (2019)
9. Yuan, B., et al.: Machine-learning-based monitoring of laser powder bed fusion. Adv Mater Technol. **3**, 12 (2018)
10. Caggiano, A., et al.: Machine learning-based image processing for on-line defect recognition in additive manufacturing. CIRP Ann. **68**(1), 451–454 (2019)
11. Qi, X., et al.: Applying neural-network-based machine learning to additive manufacturing: current applications, challenges, and future perspectives. Engineering. **5**(4), 721–729 (2019)
12. Baumgartl, H., Tomas, J., Buettner, R., Merkel, M.: A deep learning-based model for defect detection in laser-powder bed fusion using in-situ thermographic monitoring. Progr Additive Manufact. **5**, 277–285 (2020)
13. Nguyen, N.T., Ohta, A., Matsuoka, K., Suzuki, N., Maeda, Y.: Analytical solutions for transient temperature of semi-infinite body subjected to 3-D moving heat sources. Weld J. **78**, 265 (1999)
14. Baturynska, I., Semeniuta, O., Wang, K.: Application of machine learning methods to improve dimensional accuracy in additive manufacturing. In: International workshop of advanced manufacturing and automation, pp. 245–252. Springer, Singapore (2018)
15. Lee, C.W., Shin, Y.C.: Construction of fuzzy systems using least-squares method and genetic algorithm. Fuzzy Set. Syst. **137**(3), 297–323 (2003)
16. Lu, G., Kotousov, A., Siores, E.: Elementary mathematical theory of thermal stresses and fracture during welding and cutting. J. Mater. Process. Technol. **89**, 298–302 (1999)
17. Lee, C.W., Shin, Y.C.: Growing radial basis function networks using genetic algorithm and orthogonalization. Int J Innov Comput Info Contr. **5**(11), 3933–3948 (2009)
18. Manvatkar, V., De, A., DebRoy, T.: Spatial variation of melt pool geometry, peak temperature and solidification parameters during laser assisted additive manufacturing process. Mater. Sci. Technol. **31**(8), 924–930 (2015)
19. Darmadi, D.B.: Predicting temperature profile and temperature history for varied parameters of a welding process using Rosenthal's approach for semi-infinite solid. ARPN J Eng Appl Sci. **11**, 790–795 (2006)
20. Sadowski, M., Ladani, L., Brindley, W., Romano, J.: Optimizing quality of additively manufactured Inconel 718 using powder bed laser melting process. Addit. Manuf. **11**, 60–70 (2016)

Chapter 9
Thermoelastic Characterization of 3D Printed Thermoplastic Elastomers

A. Tayeb, J. -B. Le Cam, and B. Loez

Abstract This work presents an experimental characterization of the mechanical and thermomechanical properties of a soft 3D printed thermoplastic elastomer. The tested tensile specimens were obtained by Fused Deposition Modeling with a modified commercial 3D printer. Three different deposit angles with respect to the tensile loading (deposit angle of $0°$, $45°$, and $\pm 45°$) have been tested. The specimens were tested under several uniaxial tensile loadings in order to investigate both the effect of increasing strain levels from one cycle to another and the loading rates. For both tests, the full temperature field was measured by means of infrared thermography and the full kinematic field was determined with the Digital Image Correlation (DIC) technique. Results provide information on the importance of the printing strategy effect on the mechanical response, the thermal response, as well as the specimen failure.

Keywords 3D printed TPE · Digital image correlation · Thermomechanical characterization · Infrared thermography · Fused deposition modeling

9.1 Introduction

Thermoplastic elastomers (TPE) are widely used in many industrial applications; construction and civil engineering, medical and automotive applications, to name a few. Usually, TPE parts are manufactured by using mold injection, which requires several steps of prototyping and high precision in the mold manufacturing process. In this context, additive manufacturing (AM) could be a promising alternative to the mold injection process since the total manufacturing cost can decrease drastically. AM was firstly used in the prototyping phase, in the early 80^s. Its use has been overwhelmingly increasing since and it is now used in the volume production phase. Several AM technologies have been developed through the recent years. A complete overview of these technologies for polymers can be found in the review article by [1] and references therein. It should be noted, however, that these technologies have yet to be fully developed for soft materials such as thermoplastic elastomers, namely, the fused deposition modeling technique (FDM), which could be very challenging for soft materials (hardness as low as 50 shore A). In fact, these materials are not printable with commercialized 3D printing machines and a modification to the printing head has to be done.

In this work, a thermomechanical characterization of 3D printed soft thermoplastic elastomer is presented. The tensile specimens were obtained by FDM following several printing strategies. The specimens were subjected to two different mechanical loadings. The aim was to investigate the influence of the printing strategy on the mechanical, the thermomechanical, and the failure response of the specimens.

This paper is organized as follows: Section 9.2 presents the experiments carried out and the measurement techniques employed to retrieve both kinematic and thermal fields. Section 9.3 shows the obtained results, and concluding remarks close the paper.

A. Tayeb (✉) · J. -B. Le Cam
Institut de Physique de Rennes, Université de Rennes 1, Rennes, France
e-mail: adel.tayeb@univ-rennes1.fr

B. Loez
Cooper Standard, Vitré, France

© The Society for Experimental Mechanics, Inc. 2022
S. L. B. Kramer et al. (eds.), *Thermomechanics & Infrared Imaging, Inverse Problem Methodologies, Mechanics of Additive & Advanced Manufactured Materials, and Advancements in Optical Methods & Digital Image Correlation, Volume 4*, Conference Proceedings of the Society for Experimental Mechanics Series, https://doi.org/10.1007/978-3-030-86745-4_9

Fig. 9.1 Printing machine

9.2 Experiments

9.2.1 3D Printing Device

The 3D printing machine used in this study is a A4v4 model of the 3ntr company. This 3D printer is shown in Fig. 9.1. It is characterized by the following printing parameters:

- A maximum nozzle temperature of 450 °C
- A maximum printing volume of $295 \times 195 \times 190 \text{ mm}^3$
- A maximum printing platform temperature of 160 °C
- Four nozzle diameters: 0.3, 0.4, 0.6, and 0.8 mm

It should be noted that the printing machine of Fig. 9.1 in its commercialized configuration does not allow the printing of the studied material. A modification of the printing head has been realized. This modification is not presented here due to confidentiality aspects.

9.3 Material and Specimen Geometry

The material is a thermoplastic styrenic elastomer (TPE or TPE-S) which contains two types of copolymer: polystyrene polybutadiene polystyrene (SBS) and Polystyrene poly(ethylene-butylene) polystyrene with a glassy temperature of −55 °C. The specimen geometry is shown in Fig. 9.2a. It is a simple extension specimen of 105 *mm* of length, 30 *mm* width, and 2 mm thick. It is equipped with cylindrical ends of a diameter of 10 mm to avoid any slippage in the grips of the tensile machine. Three deposition angles with respect to the loading direction were studied (0°, 45°, and ± 45°) as illustrated in Fig. 9.2b.

It should be noted that the deposition angles for the first and the second configurations (angle equal to 0° and 45°) were the same for the whole specimen. However, for the ±45°, the angle change from one layer to another by 90°. The filament diameter was set to 0.4 mm as a result of a parametric study on the influence of its value on the quality of the printed sample.

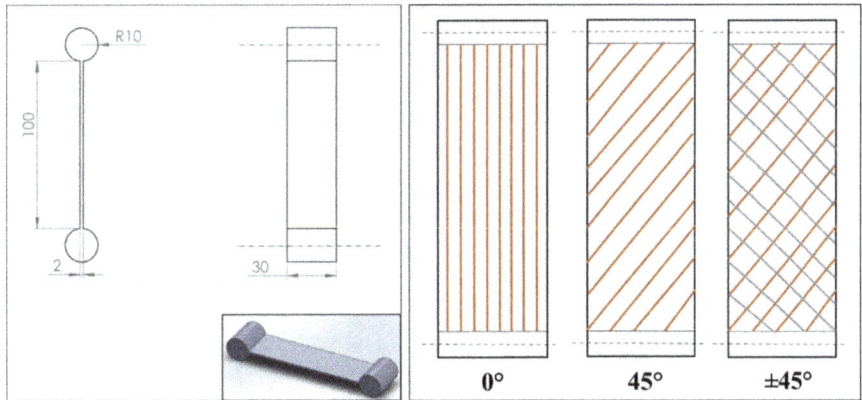

Fig. 9.2 (**a**) Specimen geometry, (**b**) deposition angles

9.4 Mechanical Loadings

The three specimens described above were subjected to two mechanical loadings; the first one consisted in applying five cycles of loading and unloading at increasing displacement levels in order to depict the influence of the deposit angles upon the thermal and mechanical responses including the hysteresis, the accommodation, and the permanent set at a moderate strain rate. The second test, on the other hand, consisted in applying one cycle of loading and unloading for a maximum displacement allowed by the current equipment (165 mm/grip) at high and low displacement rates of 500 *mm/min* and 50 *mm/min*, respectively. The aim from this experiment was to investigate the fracture behavior of the specimen and its relation with the deposition angles as well as the viscoelasticity effects from one deposition angle to another.

9.5 Experimental Setup and Measurement Techniques

The experimental setup of Fig. 9.3 is composed of a biaxial testing machine composed of four independent electrical actuators controlled by a LabVIEW program. The machine is equipped with two load cells in the two perpendicular directions with a capacity of 1000 N each. In this work, only the vertical actuator was used. Note that the applied loading for both experiments was symmetric. Therefore, the central zone of the specimen is kept motionless during the test, which is convenient for the infrared measurement since no movement compensation is needed in order to measure temperature variations of this zone. Temperature measurement was performed by using a FLIR X6540sc InSb infrared camera with the following characteristics:

- 640×512 pixels
- Wavelength range between 1.5 and 5.1 μm
- Detector pitch of 15 μm

The temperature was measured at the specimen's center (motionless part), by averaging the temperature of a zone of 5×5 pixels. This technique was used for both experiments. On the other hand, the kinematic field measurements were performed with an IDS camera equipped with a 55 mm telecentric objective. The charge-coupled device (CCD) of the camera has 1920×1200 joined pixels. The displacement field at the specimen surface was determined by using the DIC technique by using the septD software [8]. Both infrared and kinematic measurement were performed at the same rate (5 Hz). The full DIC hardware and analysis parameters are reported in Tables 9.1 and 9.2, respectively.

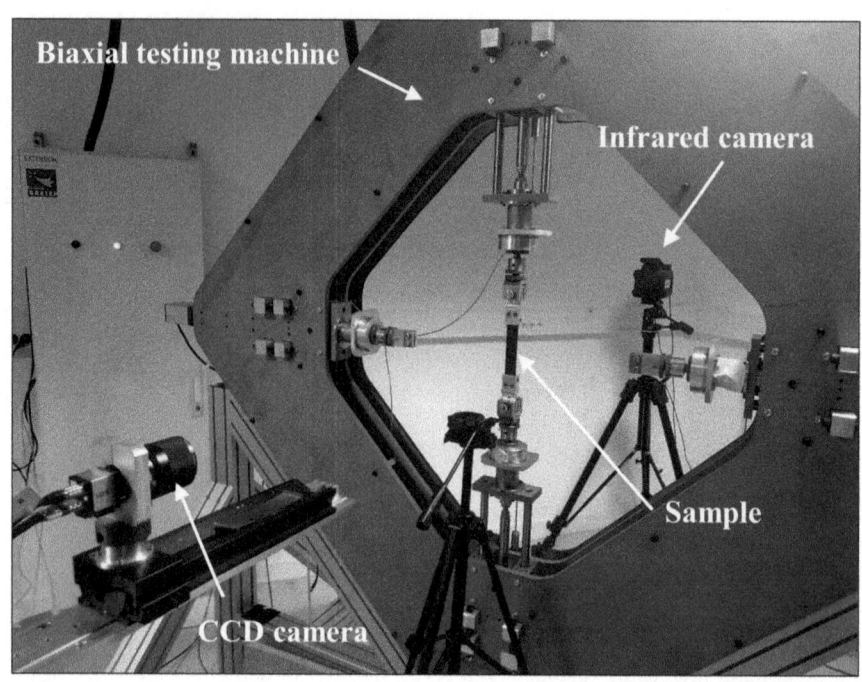

Fig. 9.3 Experimental set-up

Table 9.1 DIC hardware parameters [2]

Camera	IDS UI-3160CP Rev. 2
Image resolution	1920×1200 pixels^2
Lens	55 mm C-mount partially telecentric. Constant magnification over a range of working distances ±12.5 mm of object movement before 1% error image scale occurs
Aperture	f/5.6
Field-of-view	139.4×87.1 mm
Image scale	14 pixels/mm
Stand-off distance	1100 mm
Image acquisition rate	5 Hz
Patterning technique	White spray on black specimen
Pattern feature size (approximation)	6 pixels

Table 9.2 DIC analysis parameters [2]

DIC software	7D©
Image filtering	None
Subset size	20 pixels/1.45 mm
Step size	4 pixels/ 0.29 mm
Subset shape function	Affine
Matching criterion	Normalized cross correlation
Interpolant	Bi-cubic
Strain window	5 data points
Virtual strain gauge size	36 pixels/2.62 mm
Strain formulation	Logarithmic
Post-filtering of strains	None
Displacement noise-floor	0.036 pixels/2.6 μm
Strain noise-floor	6.1 mm/m

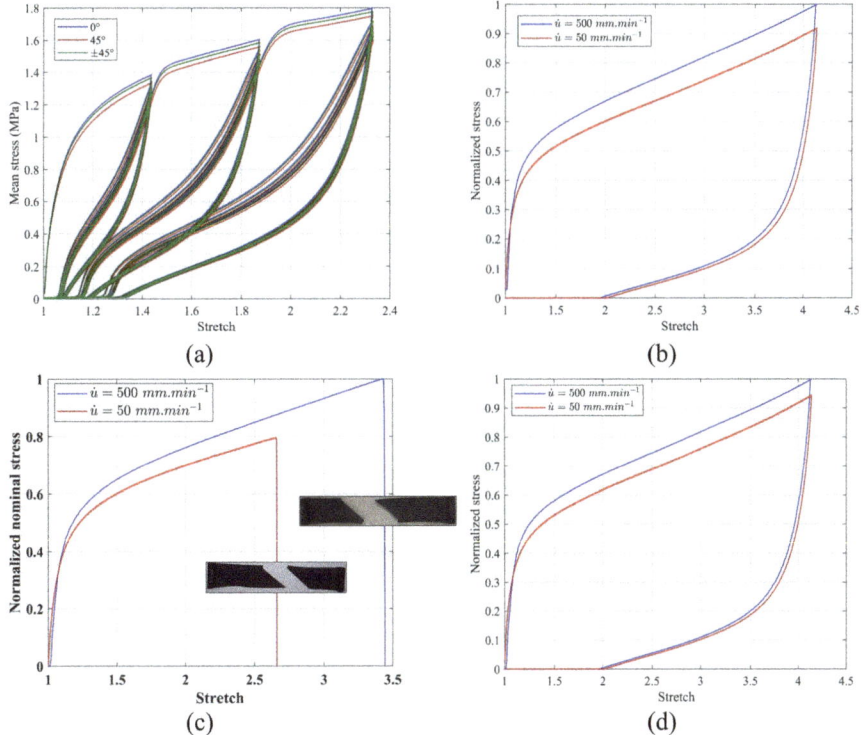

Fig. 9.4 Mechanical responses of all configurations

9.6 Results and Analysis

9.6.1 Mechanical Response

The mechanical responses of all configurations are reported in Fig. 9.4a in terms of the mean (averaged over three experiments for each configuration) nominal stress against the principal stretch. The stress-strain response of the material exhibits the same strong-nonlinearities found with filled elastomers or thermoplastic urethane [4]. Note that the mechanical responses are equivalent between all three printing configurations including the hysteresis, permanent set, and Mullin effect [7].

Figure 9.4b, c, d shows the mechanical response of the material to the second experiment for all configurations in terms of the normalized stress against the principal stretch. The normalization here was realized with respect to the high displacement rate experiment, i.e., 500 mm/min. It can be seen from all these figures that, the viscoelastic contribution to the behavior of the studied material for one order of magnitude difference in the displacement rate is important. The fracture occurred only in the case of the 45° configuration as shown in Fig. 9.4c with the same angle with respect to the loading direction.

Figure 9.5 shows the deformation in the loading direction ε_{yy} measured by using DIC for the three configurations. Figure 9.5a, b, c shows the distribution of this deformation for the 0°, 45°, and ± 45° configurations, respectively. It is found that this deformation is not homogeneous even for the zones far from the machine grips (bottom of the figures). Furthermore, the average value of this deformation is equivalent between all configurations even though it was not the case for its distribution.

9.6.2 Thermal Responses

The temperature was measured during the experiment at the specimen's center. Only results for the first experiment are reported in Fig. 9.6 for the three studied configurations. Several observations can be drawn from these results:

- At the beginning of the first loading, the temperature variation first dropped before increasing; it is the thermoelastic inversion classically observed with rubber-like materials, see [3, 5].

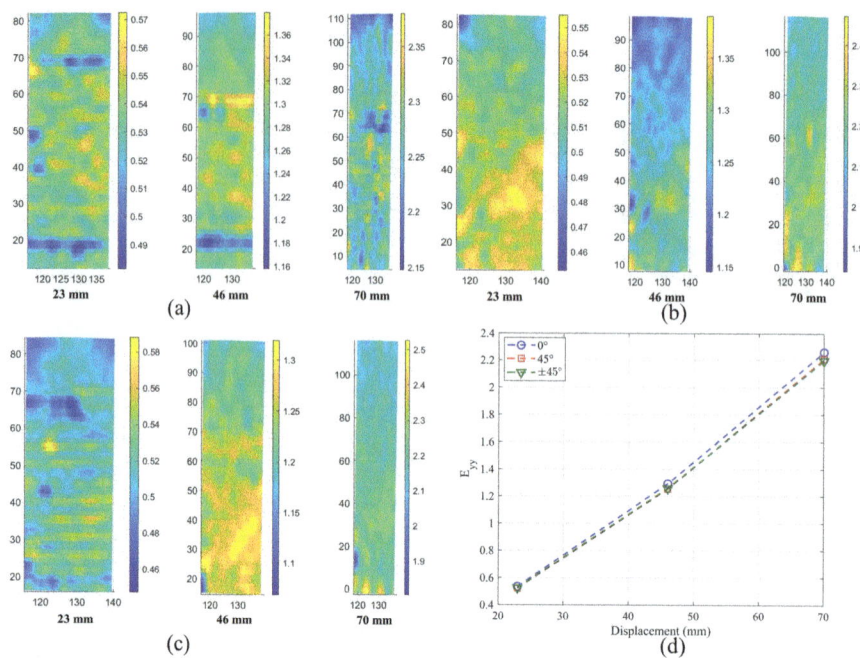

Fig. 9.5 Deformation in the loading direction

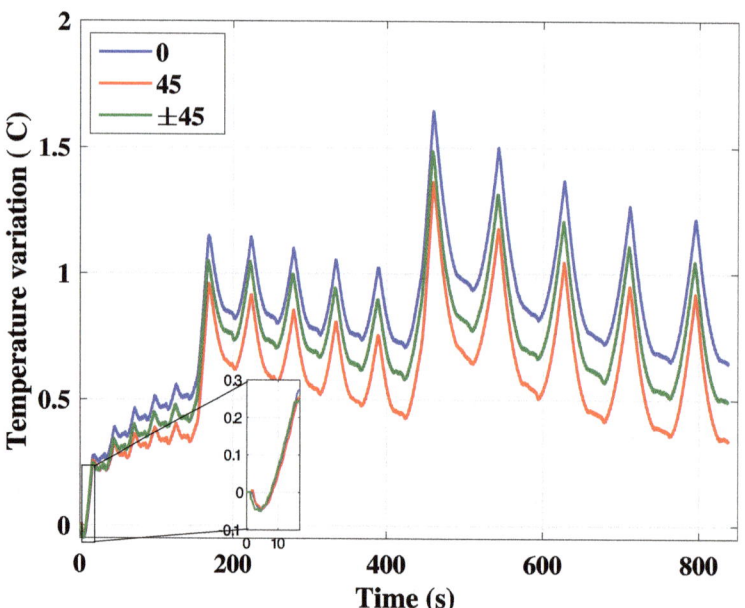

Fig. 9.6 Temperature variations versus time and printing strategy

- A significant self-heating is observed during the first loadings of each displacement level; this is explained by the additional heat production caused by the softening effects, see [6].
- Contrarily to the mechanical response, the thermal one is affected by the deposit angle. In fact the 0° configuration exhibited the maximum self-heating followed by the ±45° and 45° configurations in order. Namely, a difference of about 0.2 °C is observed since the third cycle at the first displacement level and kept increasing until the end of the test between the 0° and 45° configurations.

9.7 Conclusions

In this work, a thermoelastic characterization has been carried out for a soft thermoplastic elastomer obtained by 3D printing. The printing technique used in this work was the FDM. To this end, a modification of the printing head has been achieved in order to extend the printability to softer materials. Three printing strategies have been studied following the deposit angle with respect to the loading direction. Two mechanical loadings have been performed to the tensile specimen in order to investigate the influence of the deposit angle on the thermomechanical response of the material.

It has been showed that this angle has a slight effect on the mechanical response mainly for the time-dependent part of it, i.e., viscoelastic effects. The thermal response, however, has been affected more by the deposit angle. This result was found from the difference of the self-heating between the studied configurations. It should be noted that a small part of the results was shown in this paper, which will be fully presented during the corresponding talk.

Acknowledgments The authors thank the Cooper Standard and the Elixance companies for supporting this work, Jean-Marc Veillé and Eric Josso for fruitful discussions. The authors thank also the Région Bretagne for financially supporting this work (Région Bretagne grant IMPRIFLEX). SEM images were performed at CMEBA facility (ScanMAT, University of Rennes 1), which received a financial support from the European Union (CPER-FEDER 2007-2014).

References

1. Dizon, J.R.C., Espera Alejandro, H., Chen, Q., Advincula, R.C.: Mechanical characterization of 3D-printed polymers. Addit. Manuf. **20**, 44–67 (2019)
2. Jones, E.M.C., Iadicola, M.A.: A good practices guide for digital image correlation. Int Digital Image Corr Soc. **10**, 5 (2018)
3. Joule, J.P.: On some thermo-dynamic properties of solids. Philos Trans Roy Soc London. **149**, 91–131 (1859)
4. Lachhab, A., Robin, E., Le Cam, J.-B., Mortier, F., Tirel, Y., Canevet, F.: Thermomechanical analysis of polymeric foams subjected to cyclic loading: Anelasticity, self-heating and strain-induced crystallization. Polymer. **126**, 19–28 (2017)
5. Martinez, J.R.S., Le Cam, J.-B., Balandraud, X., Toussaint, E., Caillard, J.: Mechanisms of deformation in crystallizable natural rubber. Part 1: thermal characterization. Polymer. **54**(11), 2717–2726 (2013)
6. Martinez, J.R.S., Le Cam, J.-B., Balandraud, X., Toussaint, E., Caillard, J.: Mechanisms of deformation in crystallizable natural rubber. Part 2: Quantitative calorimetric analysis. Polymer. **54**(11), 2727–2736 (2013)
7. Mullins, L.: Effect of stretching on the properties of rubber. Rubber Chem Technol. **21**(2), 281–300 (1948)
8. Vacher, P., Dumoulin, S., Morestin, F.: Mguil-Touchal, Bidimensional strain measurement using digital images. J Mech Eng Sci. **213**, 811 (1999)

Chapter 10
Reduction of Micro-Crack in Ni-Based Superalloy IN-713LC Produced by Laser Powder Bed Fusion

M. Mohsin Raza, Hung-Yu Wang, Yu-Lung Lo, and Hong-Chuong Tran

Abstract Inconel 713LC is a Ni-based superalloy, which is known as a non-weldable alloy, subjected to severe solidification cracking during the LPBF process. In this study, a systematic optimization method was constructed to find the optimal parameter region of IN713LC in LPBF additive manufacturing. The optimization method combines the perspectives of reducing the micro-cracks and the pores in melt-pool; therefore, the optimal region of the processing map can provide workpieces with less crack and high density. The specimens were fabricated with various fabrication parameters, leading to different melt-pool sizes, shapes, and different crack density. The corresponding results shows a trend that larger the mushy zone would result in the higher susceptibility of micro-cracking. As a result, the crack density of the specimens in the optimal parameter region has the lowest crack and the high relative densiy. It is found that in the result of the tensile test, LPBF processed IN713LC specimen shows excellent mechanical properties than that in casting.

Keywords Non-weldable · Solidification cracking · Parameter optimization · Mushy zone

10.1 Introduction

Inconel 713LC (IN713LC) is a low-carbon version of nickel-based alloy IN713, which is a common casting material in the aerospace industry. Due to the chemical element composition of IN713LC, it produces a large number of cracks during welding and build-up. Therefore, IN713LC is regarded as a non-printable material [1], and there is little research on this issue. There are several causes of cracks in alloys, including solidification cracking, liquation cracking, and ductility-dip cracking (DDC) [2] inside the melting pool and the heat affected zone (HAZ) while it becomes solid from liquid phase. The full expression of liquation cracking and DDC would require a comprehensive study on microstructure, which is not inside the scope of this article. Also, it is noted that the cold crack caused by the residual stress after a part is cooled down will not be studied in this research. Usually, the cold cracks could be observed between pores [3] inside a part.

In the current study, we made great efforts for LPBF process parameters optimization, to minimize the cracking issue about the IN713-LC when processed by the LPBF. Optimization of LPBF processing parameters [4] used to manufacture samples with fewer cracks, and have made the comparison of mechanical properties with casting one is made. Also the heat-treatment effect on the mechanical properties is observed.

10.2 Map for Less Micro-Crack

Figure 10.1 shows the process map for less micro-crack, where the high crack susceptivity region is marked in magenta. The green region in Fig. 10.1 therefore corresponds the optimal processing region for LPBF in IN713LC; in other words, the single scan tracks produced with different sets of the laser power and scanning speed within this green region are free of the keyhole

M. M. Raza (✉) · H.-Y. Wang · Y.-L. Lo
Department of Mechanical Engineering, National Cheng Kung University, Tainan, Taiwan
e-mail: loyl@mail.ncku.edu.tw

H.-C. Tran
Department of Mechanical Engineering, National Cheng Kung University, Tainan, Taiwan

Department of Mechanical Engineering, Southern Taiwan University of Science and Technology, Tainan, Taiwan

S. L. B. Kramer et al. (eds.), *Thermomechanics & Infrared Imaging, Inverse Problem Methodologies, Mechanics of Additive & Advanced Manufactured Materials, and Advancements in Optical Methods & Digital Image Correlation, Volume 4*, Conference Proceedings of the Society for Experimental Mechanics Series, https://doi.org/10.1007/978-3-030-86745-4_10

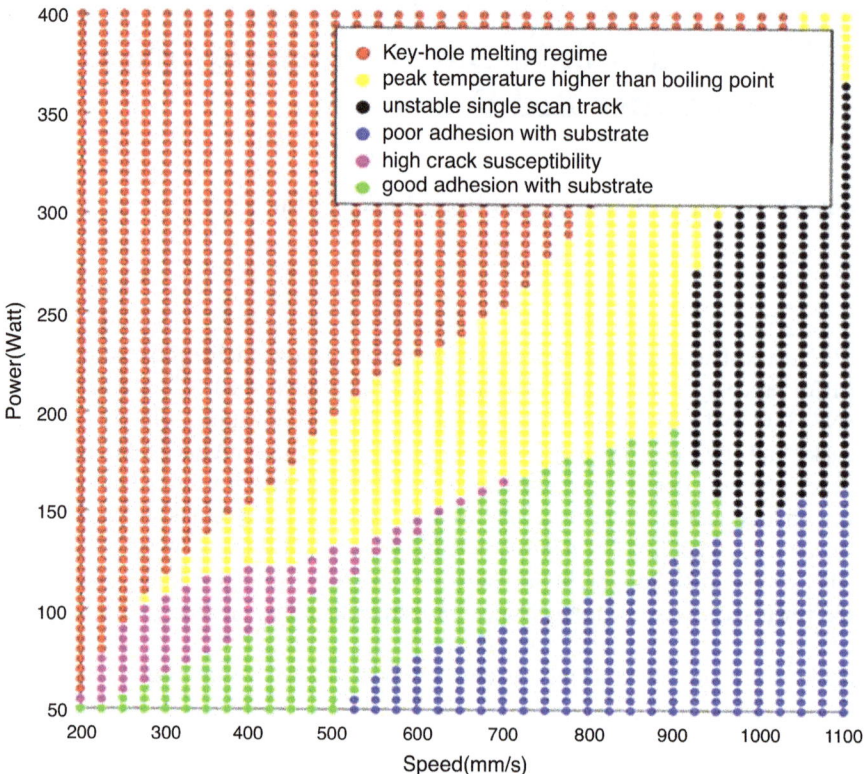

Fig. 10.1 Process map for less micro-crack; red region: keyhole melting regime; blue region: poor adhesion with substrate; black region: unstable single scan track; yellow region: peak temperature higher than boiling point; magenta region: high crack susceptivity; and green region: high density with less crack [4]

effect, having a good adhesion with the substrate, having a good stability, suffering minimal distortion, and generating less micro-crack.

10.3 Experimental Setup

All the experiments were conducted on a Tongtai AM-250 selective laser melting machine. The laser powder bed fusion system is composed of an Nd-YAG laser with a wavelength of 1064 nm, and the laser focal beam diameter is 100 μm with a Gaussian profile. The Tongtai AM-250 machine features the maximum laser power of 500 watts and the maximum laser scanning speed of 2000 mm/s. The scanning strategy was zig-zag and the scanning direction rotated 67° between layers. In order to avoid oxidation during melting, all the experiments were conducted in nitrogen atmosphere with oxygen concentration limited below 1000 ppm.

10.4 Experiments for Mechanical Properties

It is very important to understand the corresponding mechanical properties; thus, a tensile test was conducted. The laser power is fixed at 200 W and the scanning speed is fixed at 700 mm/s; only the hatching space is chosen as 80 and 100 μm. Table 10.1 presents the parameters of tensile test bar for Samples W, X, Y, and Z. In order to study the influence of heat treatment on mechanical properties, half of the specimens were taken heat treatment with 1200 °C for 2 h.

It is noted that the height of the supporter structure is 6 mm, so the total height of the specimens reaches to 21 mm as shown in Fig. 10.2. As a result, Fig. 10.3 and Table 10.2 present the tensile test results. It can be observed that the specimens with heat

Table 10.1 Parameters of tensile test bar for Samples W-Z

	Heat treatment	
Hatching space	Yes	No
80 μm	W	X
100 μm	Y	Z

Fig. 10.2 Tensile test bar fabricated by a SLM machine

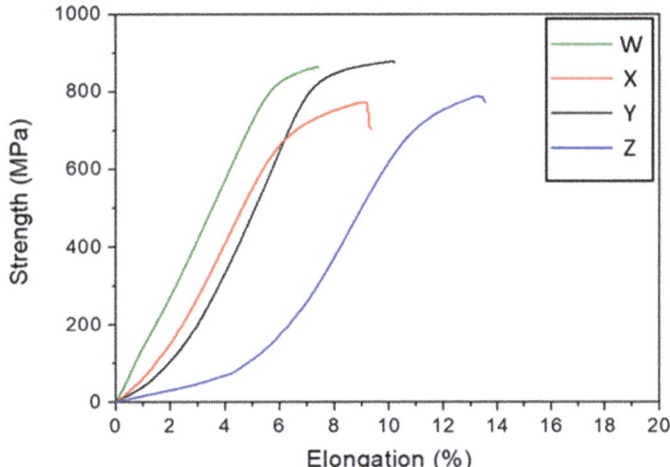

Fig. 10.3 Results of the tensile test

Table 10.2 Mechanical properties in the present study

Sample	YS (MPa)	UTS (MPa)	UE (%)	TE (%)
W	807	864	1.2	1.2
X	684	773	1.8	2.7
Y	829	878	2.1	2.1
Z	749	789	1.1	1.4

treatment have the greater yield strength and ultimate tensile strength; however the elongation is lower. The specimens with 100 μm hatching space have the greater elongation. It is concluded that all of the results have better yield strength and ultimate tensile strength, but the elongation is lower than those of as-cast IN713LC in AMS 5377 in Table 10.3 [5].

Table 10.3 Tensile properties of as-cast IN713LC in AMS 5377 [5]

	T (°C)	YS (MPa)	UTS (MPa)	Elongation (%)
As-cast	20	≥689	≥758	≥5

10.5   Conclusions

The significant results and achievements in this study are summarized as:

IN713LC is successfully build using the processing map build by our group in already submitted paper [4]. An experiment for investigating mechanical properties of a 3D part was conducted, and the result of the tensile test has the better strength, but the elongation is lower than that of as-cast specimen in AMS 5377.

Acknowledgments   The authors gratefully acknowledge the financial support provided to this study by the Ministry of Science and Technology of Taiwan under Grant No. MOST 107-2218-E-006-051. The study was also supported in part by the funding provided to the Intelligent Manufacturing Research Center (iMRC) at National Cheng Kung University (NCKU) by the Ministry of Education, Taiwan, Headquarters of University Advancement. The assistance provided by Prof. Fei-Yi Hung at NCKU in conducting the heat treatment and tensile testing elements of the present study is much appreciated.

References

1. Koren, A., Roman, M., Weisshaus, I., Kaufman, A.: Improving the weldability of Ni-base superalloy 713 C. Weld J. **61**(11), 348–351 (1982)
2. Zhang, X.Q., Chen, H.B., Xu, L.M., Xu, J.J., Ren, X.K., Chen, X.Q.: Cracking mechanism and susceptibility of laser melting deposited Inconel 738 superalloy. Mater. Des. **183**, 14 (2019)
3. Guo, C., et al.: Effect of processing parameters on surface roughness, porosity and cracking of as-built IN738LC parts fabricated by laser powder bed fusion. J. Mater. Process. Technol. **285**, 5 (2020)
4. Wang, L., Raza, Tran: Systematic approach for reducing micro-crack formation in Inconel 713LC components fabricated by laser powder bed fusion. Rapid Prototyping J. **27**(8), 1548–1561 (2021)
5. Heaney, D.F.: Handbook of Metal Injection Molding. Woodhead Publishing, New York (2018)

Chapter 11
Analysis of the Thermomechanical Behaviour of SMP in Equi-Biaxial Condition by Means of Hydraulic Bulge Test

Mattia Coccia, Attilio Lattanzi, Gianluca Chiappini, Marco Sasso, and Marco Rossi

Abstract In the last few decades, the shape memory polymers have gained growing interest and relevance in many application fields, which include textiles, aerospace, biomedical devices, etc. SMP are materials with stimuli-sensitive switches that are able to change their geometry from a primary shape to a secondary shape—and vice versa—in response to external stimuli. This phenomenon is called shape memory effect and, generally, is triggered by heat.

The heat stimulation, in fact, alters the internal structure of the polymer: by exceeding the glass transition temperature T_g, it is possible to program the shape of the component. Then, by cooling the material below T_g and imposing a fixed deformation, the polymer can reach its temporary shape. The original shape can be recovered by heating again the material above T_g without any constraints.

Recent research works are mostly focused on thermomechanical uniaxial characterization of these materials, aimed to obtain the material main memory effect parameters (namely the Young modulus above/under glass transition temperature and shape fixed/recovery ratio). In this work the authors propose a non-conventional characterization approach for the investigation of SMP behaviour under equi-biaxial stress state. In particular, the main idea is to carry out the hydraulic bulge test with a thermomechanical cycle on thin sheets of thermoplastic polyurethane shape memory polymer (TPU-SMP).

Full-field measurements on the specimen surface are used for retrieving the material shape, this latter by employing the Digital Image Correlation (DIC) technique. The proposed configuration makes the test suitable for determining the biaxial-stress strain curve and the thermomechanical cycle, also providing data for inverse calibration methods such as the Virtual Fields Method (VFM) and the Finite Element Model Updating (FEMU).

Keywords Shape memory polymers · Shape memory effect · Thermomechanical test · Hydraulic bulge test · Equi-biaxial test

11.1 Introduction

SMP materials are thermomechanically functional smart materials [1]. Shape memory polymers (SMP) have gained growing worldwide interest and relevance in many application fields, such as actuators [2], robotics [3], aerospace industry [4], biomedical applications [5], such as sutures [6], ureteral stent [7], device for stroke patients [8], orthodontic [9] and vascular stent [10]. Thanks to their high recoverable strains, low cost, stimuli-sensitiveness, biodegradability, and biocompatibility, at present SMP materials are widely employed for mechanical analysis [11].

In particular, the SMPs are polymers that undergo a very strong shape memory effect (SME), leading the material to come back, from a deformed temporary shape, to its original permanent shape under an external stimulus, such as the heat, magnetism, electricity, light, and some specific chemicals [12]. Generally depending on the application fields, the use of smart materials, along with the SME, requires a specific characterization based on their chemical and mechanical properties. In order to mechanically characterize the SME, a thermomechanical cycle is performed to determine two parameters, one to ensure the efficiency of the material in maintaining the temporary shape, while the other to recover such shape [13]. The thermomechanical cycle, summarized in Fig. 11.1, is composed of five steps (i.e. heating, deformation, cooling, unloading and reheating) obtained by controlling three main variables: the temperature of the test, the stress and strain generated on the

M. Coccia (✉) · A. Lattanzi · M. Sasso · M. Rossi
Department of Industrial Engineering and Mathematical Sciences, Università Politecnica delle Marche, Ancona, Italy
e-mail: m.sasso@univpm.it

G. Chiappini
Università degli Studi eCampus, Novedrate, Italy

© The Society for Experimental Mechanics, Inc. 2022
S. L. B. Kramer et al. (eds.), *Thermomechanics & Infrared Imaging, Inverse Problem Methodologies, Mechanics of Additive & Advanced Manufactured Materials, and Advancements in Optical Methods & Digital Image Correlation, Volume 4*, Conference Proceedings of the Society for Experimental Mechanics Series, https://doi.org/10.1007/978-3-030-86745-4_11

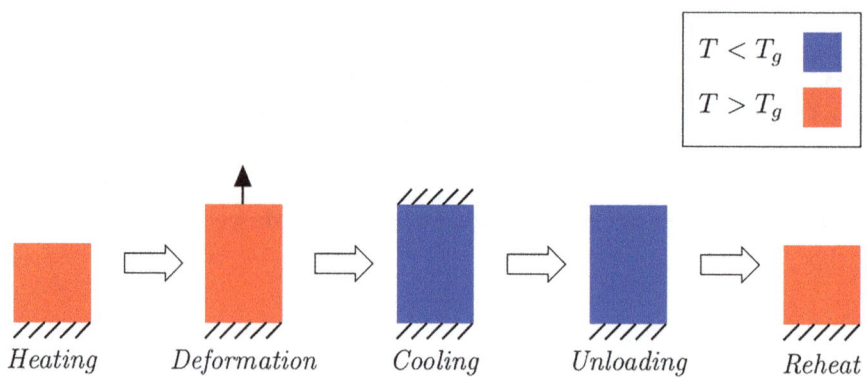

Fig. 11.1 Schematic of shape memory effect

specimen [14, 15]. In detail, during the first and second step, the specimen is strained to a prescribed value, ε_m, at a temperature above the glass transition temperature (T_g). Then in the third step, the specimen is cooled at the room temperature (below T_g), maintaining constant the deformation obtained in the second phase. In this way, a new shape is impressed and learned to the specimen. In the fourth step, the specimen is unloaded at room temperature and the stress is reduced to zero, for assessing the strain ε_u as output. Once the step terminates, the shape fixity ratio R_f is computed as:

$$R_f = \frac{\varepsilon_u}{\varepsilon_m} * 100 \tag{11.1}$$

The last step consists in heating again above T_g under no load condition, where the irrecoverable strain ε_{ir} remains at the heating finish point and the shape recovery ratio R_r can be found as:

$$R_r = \frac{\varepsilon_m - \varepsilon_{ir}}{\varepsilon_m} * 100 \tag{11.2}$$

In the literature, the characterization of shape memory materials has been mainly conducted with uniaxial test [16], including only few experimental data on shape memory membranes [17]. Hence, the need to investigate the behaviour of the SMPs above the T_g has led us to develop a system to perform a specific balanced biaxial test by means of hydraulic bulge test. This kind of test has been often utilized to perform biaxial tensile tests on rubbers and polymers at room temperature [18]. Hydraulic bulge test in a temperature-controlled environment has been performed by Lee et al. in [19] on advanced high strength steels under warm conditions (test temperature below 100 °C); in this case, the experimental setup employed a LVDT combined to an extensometer in order to provide a simultaneous measurement of the bulge height and the in-plane elongation at the pole of the specimen simultaneously. In [20], an high-temperature bulge test was developed for the elastic and plastic behaviour of thin films, employing a scanning laser beam for retrieving the bulge dome height, while a temperature-dependent bulge test for elastomers assisted with a 3D Digital Image Correlation (DIC) system was presented in [21]. The application of the Digital Image Correlation technique [22], in fact, allows to resolve the full-field deformation history on the specimen surface, offering the capabilities to include the observation of gradients and inhomogeneities associated with the material behaviour or loading conditions. Nowadays DIC has become a widespread method applied in many engineering problems such as dynamic tests on the Split Hopkinson Bar, [23], or even to reconstruct the volume deformation of flat and round tensile specimens from 3D surface DIC measurements [24].

In this work we describe a test bench where the specimen, in the shape of a thin disc, was tested. The specimen deformation has been recorded by means of a stereo camera system, applying DIC in order to measure the specimen displacement according to the applied pressure load and temperature cycle.

11.2 Experimental Test on TPU-SMP

The experimental activity was focused on the analysis of the Shape Memory Thermoplastic Polyurethane (TPU-SMP) behaviour, in both uniaxial tensile and balanced biaxial stress states. In both cases, the mechanical tests were conducted by controlling the overall temperature in order to activate the shape memory nature of the material, employing two Pixelink® BU371F cameras (1280×1024 pixel^2 8-bit sensor) for the stereo DIC measurement. Digital image correlation was performed

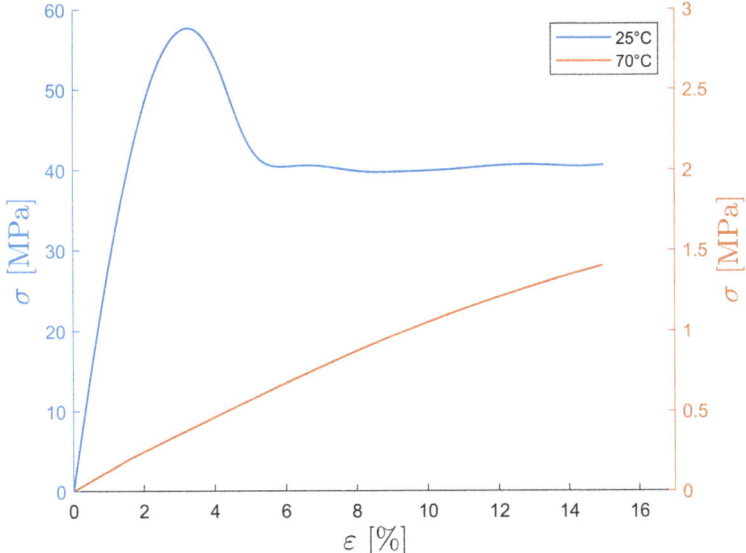

Fig. 11.2 Experimental Stress and Strain curves at two different temperatures

Table 11.1 Mechanical properties of tested material

T [°C]	E [MPa]	ν
25	2460	0.40
70	13	0.48

by commercial software MatchID 2020, using a fixed subset size (31 pixel) and step size (5 pixel). The specimens were obtained from a 0.40 mm thick laminated foil supplied by MAIP GROUP®. In particular, both the standard uniaxial specimens (ISO 527-2) [25] and the circular bulge test specimens (with a diameter of 140 mm) were cut by means of a punching machine.

Since the T_g of Shape Memory Thermoplastic Polyurethane is 60 °C, during our tests the specimens were heated at 70 °C, for surely excluding the phase of glass transition. In order to simulate the bulge test, a uniaxial tensile test was performed at different temperatures, above and below the T_g. Differently, in the bulge test a thermomechanical cycle as shown in Fig. 11.1 was executed. The following sections report the experimental results.

11.3 Uniaxial Tensile Test under Warm Conditions

Figure 11.2 describes the experimental stress and strain curves retrieved from the uniaxial tests on the SMP-TPU at room temperature (25 °C) and above the glass transition temperature of the material, namely 70 °C. It is worth to note that at 25 °C the material exhibits the typical behaviour associated to a glassy polymer [26], characterized by an initial elastic phase reaching a maximum point, referred as yield, followed by a constant stress section in the plastic regime. Once the temperature exceeds the T_g (red line in

Fig. 11.2), due to the molecular switches [27] of the material, the TPU shows a strongly different behaviour in the rubbery state, with a significant alteration of the Young modulus. The measured elastic properties of the TPU in both glassy and rubbery state are listed in Table 11.1.

11.4 Hydraulic Bulge Test

The thermomechanical tests in balanced biaxal stress condition were conducted using a hydraulic bulge test machine. This machine was placed inside a climatic chamber to maintain a constant temperature of the sample. In particular, the temperature inside the climatic chamber was regulated by employing a system of resistors and Vortec Air Guns, as shown in Fig. 11.3. Furthermore, to ensure a homogeneous heating on the specimen during the experimental tests, the water temperature was

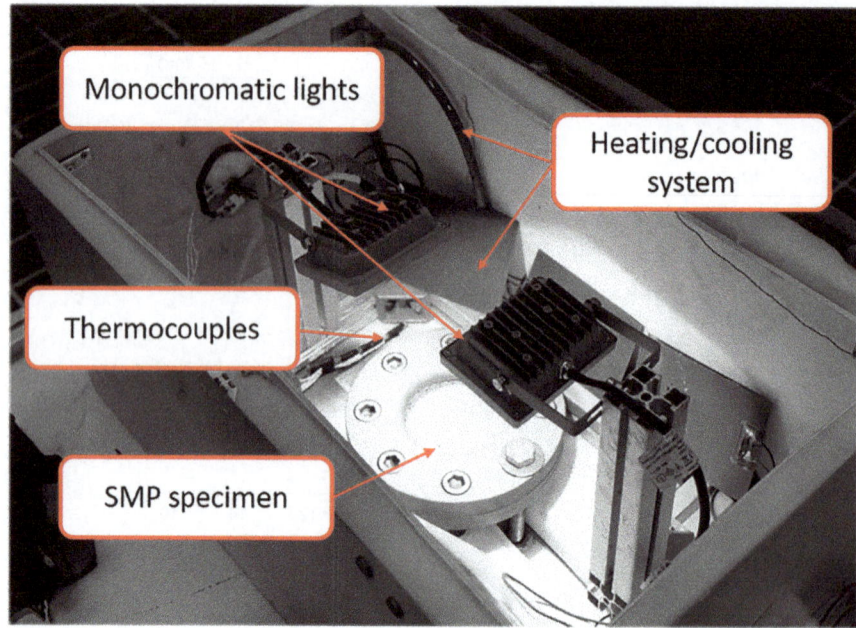

Fig. 11.3 Setup of the thermomechanical bulge test

maintained above T_g (i.e. 70 °C), for increasing the pressure. Mean temperature was monitored and recorded, by 3 K type thermocouples suspended on the specimen and in the climatic chamber, while the fluid pressure is measured with a pressure transducer (accuracy of 0.01 bar).

The height of bulge vs. time curves obtained from the tests is depicted in Fig. 11.4b, where it is possible to appreciate the different phases of the test. According to the steps describing the thermomechanical cycle (as mentioned in the introduction), the specimen was uniformly heated, and then the load was applied. Figure 11.4a shows the DIC measured displacement of the bulge in the out-of-plane direction at the end of loading phase (with a maximum of the dome at 16.70 mm and pressure of 0.3 bar). From this phase, the achieved pressure was maintained throughout the cooling step, until the temperature decreased below the T_g. After this, the pressure was unloaded to reach the zero-stress condition. The corresponding height of dome represented the amount of the deformation, which the material can store from the previous step, resulting in a fixity ratio $R_f = 98.93\%$. Finally heating again without any constraints, the specimen reacted by recovering its initial shape. However, an imperfect recovery of the shape was clear due to different factors, such as the temperature and the deformation reached in the loading phase [14]. The recovery ratio determined at the end of thermomechanical cycle was $R_r = 53.11\%$.

The bulge height was tracked over all the experimental test, for determining the fixity and recovery ratio, according to Eqs. (11.1) and (11.2). Both parameters, together with young and Poisson determined in the uniaxial tensile test, are essential to describe the non-ideal SME behaviour in numerical simulation.

11.5   Numerical Validation

A three-dimensional finite element model was developed to simulate the thermomechanical tests in the five steps of the test (as shown in Fig. 11.1) and compare with the experimental data obtained. The analysis was carried out by using the finite element simulation analysis software ABAQUS/Standard. Here, the material model employed to describe the shape memory behaviour of the TPU is the one proposed by Boatti et al. in [28], implemented in the FE code by means of a material user subroutine (UMAT). The subroutine exploits a classical implicit return-mapping algorithm to calculate the glassy phase plasticity; however other non-implicit approaches can be used, as the one presented in [29].

Basically, the numerical model of the bulge test is composed of two main parts: the specimen and the upper die. In particular, the specimen was modelled exploiting ¼ symmetry of the problem and using 8-node full-integration brick elements (C3D8). In order to enhance the computational efficiency of the simulation, only the upper die was introduced in the simulation, this latter modelled as analytical rigid surface. The tangential behaviour was described by assuming a frictional contact between the upper die and the specimen surface, with a static friction coefficient of 0.35 (Fig. 11.5).

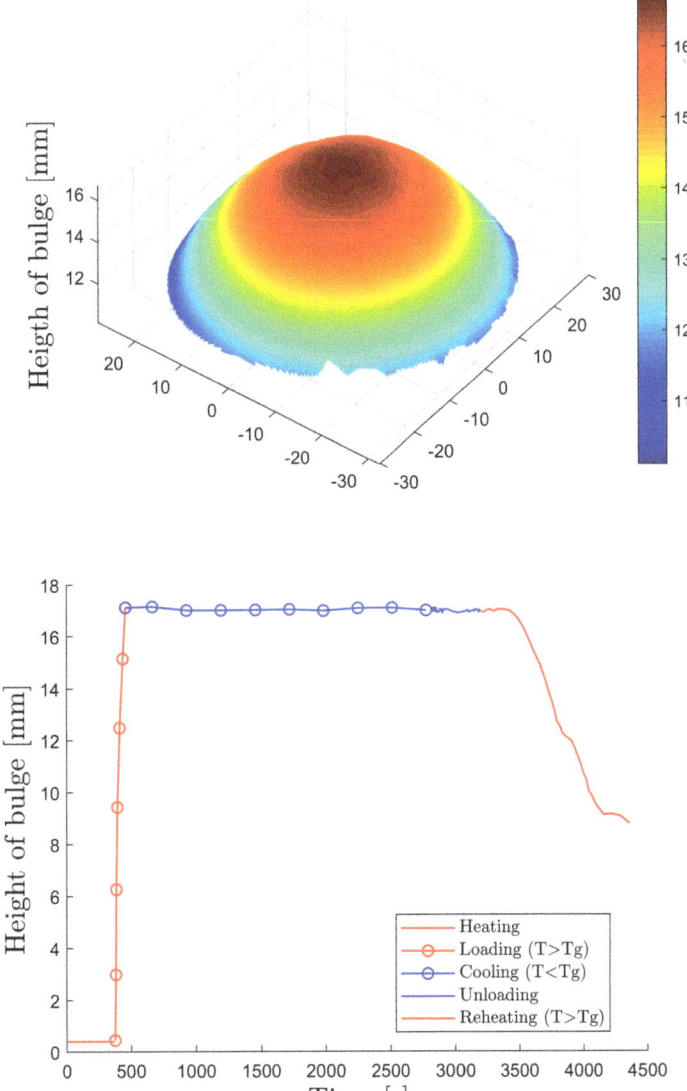

Fig. 11.4 Measured bulge height during the test. (a) Full-field map at the maximum pressure load, (b) maximum dome height during the five steps of the test

Fig. 11.5 Schematic view
of the finite element model
of the Bulge Test

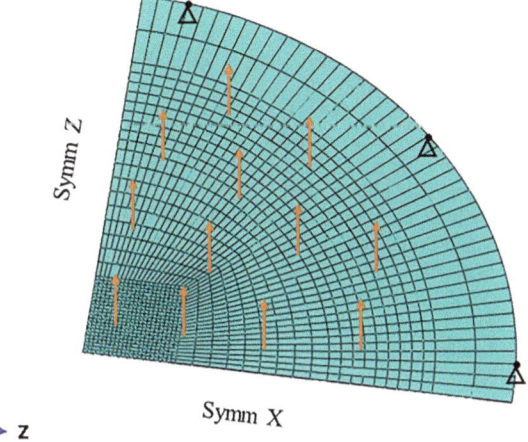

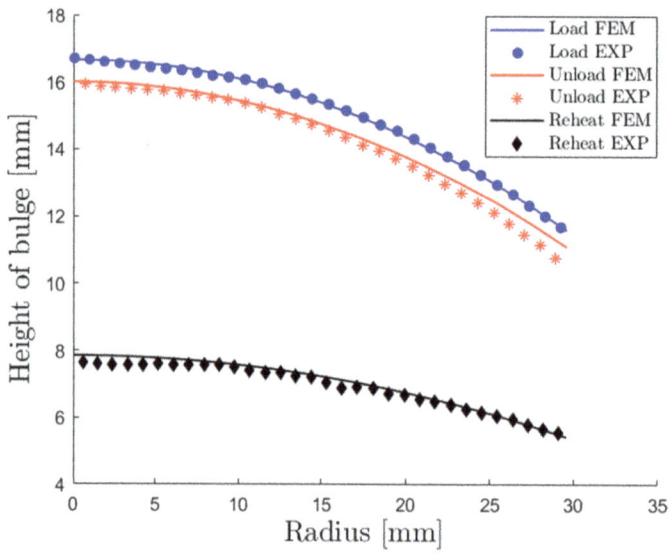

Fig. 11.6 Comparison of experimental and FEM data

Figure 11.6 shows the height of bulge vs. radius curves obtained from the comparison between the extrapolated experimental results with the analytical model. In particular the comparison of the loading, unloading and heating phases was considered.

11.6   Conclusion

In this paper, a bulge test on shape memory polymer polyurethane have been presented. Shape memory polymers with glass transition temperature at 60 °C have been tested. The tensile and the bulge tests were conducted at specified temperatures, controlled with a climatic chamber. The collected experimental data allowed to evaluate the principal parameters that characterize the shape memory effect of the material. Moreover, the maximum dome height during the five steps has been used to calibrate the analytical model, based on phases-transition of the material (glassy and rubbery state).

The capability of the analytical model of considering the fixity and recovery ratios provides an appropriate estimation of curvature between the experimental and the analytical curves, during thermomechanical cycle.

References

1. Sun, L., Huang, W.M., Ding, Z., Zhao, Y., Wang, C.C., Purnawali, H., Tang, C.: Stimulus-responsive shape memory materials: a review. Mater. Des. **33**, 577–640 (2012)
2. Chen, T., Bilal, O.R., Lang, R., Daraio, C., Shea, K.: Autonomous deployment of a solar panel using elastic origami and distributed shape-memory-polymer actuators. Phys Rev Appl. **11**, 064069 (2019)
3. Chen, T., Bilal, O.R., Lang, R., Daraio, C., Shea, K.: Harnessing bistability for directional propulsion of soft, untethered robots. PNAS. **115**, 5698–5702 (2018)
4. Keihl M.M., Bortolin, R.S., Sanders B.: Mechanical properties of shape memory polymers for morphing aircraft applications. SPIE Conference: Smart Structures and Materials (2005)
5. Melocchi, A., Uboldi, M., Inverardi, N., Briatico-Vangosa, F., Baldi, F., Pandini, S., Scalet, G., Auricchio, F., Cerea, M., Foppoli, A., Maroni, A., Zema, L., Gazzaniga, A.: Expandable drug delivery system for gastric retention based on shape memory polymers: development via 4d printing and extrusion. Int. J. Pharm. **571**, 118700 (2019)
6. Lendlein, A., Langer, R.: Biodegradable, elastic shape-memory polymers for potential biomedical applications. Science. **296**, 1673–1676 (2002)
7. Neffe, A.T., Hanh, B.D., Steuer, S., Lendlein, A.: Polymer networks combining controlled drug release, biodegradation, and shape memory capability. Adv. Mater. **21**, 3394–3398 (2009)
8. Metzger, M.F., Wilson, T.S., Schumann, D.: Mechanical properties of mechanical actuator for treating ischemic stroke. Biomed. Microdevices. **4**, 89–96 (2002)
9. Jung, Y.C., Cho, J.W.: Application of shape memory polyurethane in orthodontic. J. Mater. Sci. Mater. Med. **21**, 2881–2886 (2008)

10. Yakacki, C.M., Shandasa, R., Lanningb, C., Recha, B., Ecksteina, A., Gall, K.: Unconstrained recovery characterization of shape-memory polymer networks for cardiovascular applications. Biomaterials. **28**, 2255–2263 (2007)
11. Liu, C., Qin, H., Mather, P.T.: Review of progress in shape-memory polymers. J. Mater. Chem. **17**, 1543 (2007)
12. Leng, J., Lan, X., Liu, Y., Du, S.: Shape memory polymer and their composites: stimulus methods and applications. Prog. Mater. Sci. **56**, 1077–1135 (2011)
13. Imran, K.M., Zagho, M.M., Shakoor, R.A.: A brief overview of shape memory effect in thermoplastic polymers. Polymer Compos Mater. **5**, 281–301 (2017)
14. Bilim, A., Farhan, G., Greg, K.: Thermomechanical characterization of shape memory polymers. J. Intell. Mater. Syst. Struct. **20**, 87–95 (2008)
15. Tobushi, H., Matsui, R., Takeda, K., Hayashi, S.: Mechanical testing of shape memory polymers for biomedical applications. In: Shape Memory Polymers for Biomedical Applications, pp. 65–75. Elsevier, Amsterdam (2015)
16. Liu, Y., Gall, K., Dunn, M.L., Greenberg, A.R., Diani, J.: Thermomechanics of shape memory polymers: uniaxial experiments and constitutive modeling. Int. J. Plast. **22**, 279–313 (2006)
17. Poilane, C., Delobelle, P., Lexcellent, C., Hayashi, S., Tobushi, H.: Analysis of the mechanical behavior of shape memory polymer membranes by nanoindentation, bulging and point membrane deflection tests. Thin Solid Films. **379**, 156–165 (2000)
18. Sasso, M., Amodio, D.: Development of a biaxial stretching machine for rubbers by optical methods. Proc SEM Annual Conf Expos Exp Appl Mech. **3**, 1161–1171 (2006)
19. Lee, J.-Y., Xu, L., Barlat, F., Wagoner, R.H., Lee, M.-G.: Balanced biaxial testing of advanced high strength steels in warm conditions. Exp. Mech. **53**, 1681–1692 (2013)
20. Kalkman, A.J., Verbruggen, A.H., Janssen, G.C.A.M.: High-temperature bulge-test setup for mechanical testing of free-standing thin films. Rev. Sci. Instrum. **74**, 1383–1386 (2003)
21. Cakmak, U.D., Kallai, I., Major, Z.: Temperature dependent bulge test for elastomers. Mech. Res. Commun. **60**, 27–32 (2014)
22. Sutton, M., Orteu, J., Schreier, H.W.: Image Correlation for Shape, Motion and Deformation Measurements. Springer, New York (2009)
23. Sasso, M., Mancini, E., Chiappini, G.: Application of DIC to static and dynamic testing of agglomerated Cork material. Exp. Mech. **58**(7), 1017–1033 (2018)
24. Rossi, M., Cortese, L., Genovese, K., Lattanzi, A., Nalli, F., Pierron, F.: Evaluation of volume deformation from surface DIC measurements. Exp. Mech. **58**, 1181–1194 (2018)
25. Daniele, V., Macera, L., Taglieri, G., Di Giambattista, A., Spagnoli, G., Massaria, A., Messori, M., Quagliarini, E., Chiappini, G., Campanella, V., Mummolo, S., Marchetti, E., Marzo, G., Quinzi, V.: Thermoplastic disks used for commercial orthodontic aligners: complete physico-chemical and mechanical characterization. Materials. **13**(10), 2386 (2020)
26. Farotti, E., Natalini, M.: Injection molding. Influence of process parameters on mechanical properties of polypropylene polymer. A first study. Procedia Struct Integr. **8**, 256–264 (2018)
27. Lendlein, A., Kelch, S.: Shape-memory polymers. Angewandte Chemie Int. **41**, 2034 (2002)
28. Boatti, E., Scalet, G., Auricchio, F.: A three-dimensional finite-strain phenomenological model for shape-memory polymers: formulation, numerical simulations, and comparison with experimental data. Int. J. Plast. **83**, 153–177 (2016)
29. Rossi, M., Lattanzi, A., Cortese, L., Amodio, D.: An approximated computational method for fast stress reconstruction in large strain plasticity. Int. J. Numer. Methods Eng. **121**, 3048–3065 (2020)

Chapter 12
Inverse Identification of the Post-Necking Behavior of Metal Samples Produced with Additive Manufacturing

Marco Rossi, Gianluca Chiappini, Emanuele Farotti, and Mattia Utzeri

Abstract Additive manufacturing (AM) is undoubtedly the fastest-growing technology in the field of component productions. In particular, metal AM is rapidly emerging thanks to its enormous potentiality in manufacturing components with complex shapes and high structural performances. Although the precision and the rapidity of the production process is continuously improving, the performance of the final components in terms of crashworthiness and mechanical properties is still under evaluation. In this work, several specimens were manufactured using metal AM, in particular the Selective Laser Melting (SLM) method was employed. The used material is steel. All specimens have the same dimensions, but they were created using different paths of the laser source with respect to the metal powder layers during the AM process. Afterwards, the specimens were subjected to tensile test and the deformation field was measured using Digital image Correlation. The post-necking behavior as well as the anisotropy were evaluated using the Virtual Fields Method. It turned out that the laser paths used during the forming process have an impact on the plastic flow at large deformation up to the final fracture. The variance of the mechanical properties and the experimental uncertainties are discussed thoroughly.

Keywords Additive manufacturing · Post-necking behavior · Plasticity · DIC · Inverse methods

12.1 Introduction

Nowadays, Additive Manufacturing (AM) technologies are already widely used for cutting-edge applications in the aerospace, medical, defense, and automotive sectors [1, 2]. However, 3D printed components are more and more employed in common industrial applications too, thanks to the recent technological advancements that considerably reduced the production costs and times [3]. One of the most popular metal AM methods is Powder Bed Fusion (PBF), which allows for better surface quality and higher dimensional accuracy with respect to other metal AM technologies, as Binder Jetting (BJ) and Direct Energy Deposition (DED) [4]. According to ISO/ASTM 52900 the Powder Bed Fusion (PBF) is the additive manufacturing process in which thermal energy selectively fuses regions of a powder bed. The PBF techniques can be further classified depending on the thermal energy source. For instance, Electron Beam Melting (EBM) is electron beam-based while Selective Laser Melting (SLM) is laser-based. Both energy sources melt or sinter the metal powder on the building platform. Once the layer is complete, the build table lowers of the layer's thickness. Then, a blade (or roller) recoats the building platform with new powder. The process takes place till the component is entirely manufactured. During the printing, inert gas flows above the platform to protect the melting pool.

The process parameters are related to laser, scan pattern, powder, and temperature; each one can affect the material with discontinuities and porosities [5, 6]. Besides, the recoating transition and the inert gas flowing can pollute the clean powders on the surface with melt residues of the previous layers [7, 8]. The goal of 100% density material, under all melting conditions, is the opposite of the goal to speed rates. High solidification rates may result in metastable microstructures and material textures [9]. Therefore, discontinuities, flaws, porosities, and microstructures of SLM parts can lead to unpredictable properties, decrease of performance, and strong anisotropy of the materials [10, 11]. The crashworthiness and damage assessment of AM components is still scarcely addressed [12], although an extensive literature is available for metals

M. Rossi (✉) · E. Farotti · M. Utzeri
Department of Industrial Engineering and Mathematical Sciences, Università Politecnica delle Marche, via Brecce Bianche, Ancona, Italy
e-mail: m.rossi@staff.univpm.it

G. Chiappini
Università degli Studi eCampus, Novedrate, Como, Italy

© The Society for Experimental Mechanics, Inc. 2022
S. L. B. Kramer et al. (eds.), *Thermomechanics & Infrared Imaging, Inverse Problem Methodologies, Mechanics of Additive & Advanced Manufactured Materials, and Advancements in Optical Methods & Digital Image Correlation, Volume 4*, Conference Proceedings of the Society for Experimental Mechanics Series, https://doi.org/10.1007/978-3-030-86745-4_12

produced with standard methods. In order to understand and predict the fracture process, the first step is characterizing the constitutive behavior of the material at large deformations up to the final fracture [13]. Dealing with tensile tests, this means evaluating the properties of the material in the post-necking region. In this paper, the post-necking behavior of specimens manufactured with the SLM technology was characterized using the Virtual Fields Method [14]. The strain field during the test up to the final fracture was evaluated through Digital Image Correlation (DIC). Both VFM and DIC are well-established techniques that, in recent years, proved to be effective in a large range of applications, including complex materials, dynamic behavior, etc. [15, 16].

12.2   Experiments

A series of specimen was manufactured using the SLM method. The geometry of all specimens is the same, i.e., the classical dog-bone shape prescribed by the norms, but they are manufactured according to different orientations with respect to the laser scan direction and the metal powder layers. Figure 12.1 shows an example of the orientations that can be obtained that, here, will be named using two angles separated by a dash. The first angle represents the inclination with respect to the platform (i.e., metal powder layers); the second angle is the orientation with respect to the laser scan directions. Clearly, a large number of directions should be tested to have a full insight of the material behavior; however, in this preliminary study only two directions were tested, 0°–45° and 90°–0°, respectively. The two directions are highlighted in red in Fig. 12.1. The intent here is to investigate two situations that are very different in terms of AM process, and evaluate how much the AM process has an impact on the material properties. An extensive study of all directions is postponed to future work.

The tensile tests were performed using a standard Zwick/Roell® Z050 machine, equipped with a 50kN load cell. The specimens' surface was suitably prepared using black and white spray paint to generate the speckle pattern for DIC measurements. The crosshead speed was set to 2 mm/min and images were acquired every 2 s; the system was synchronized to match the load value measured during the test with the corresponding image. A single digital camera was used to perform 2D DIC and a high voltage lamp was used to have a proper illumination and improve the quality of the images. The experimental setup is illustrated in Fig. 12.2. It should be noted here that, in the present case, the analysis is focused on the post-necking regime where the measured strain is large. Under this condition, possible errors due to out-of-plane movements [17] can be neglected and, furthermore, in order to reduce their impact, a long distance was set between the camera and field of view, see Fig. 12.3.

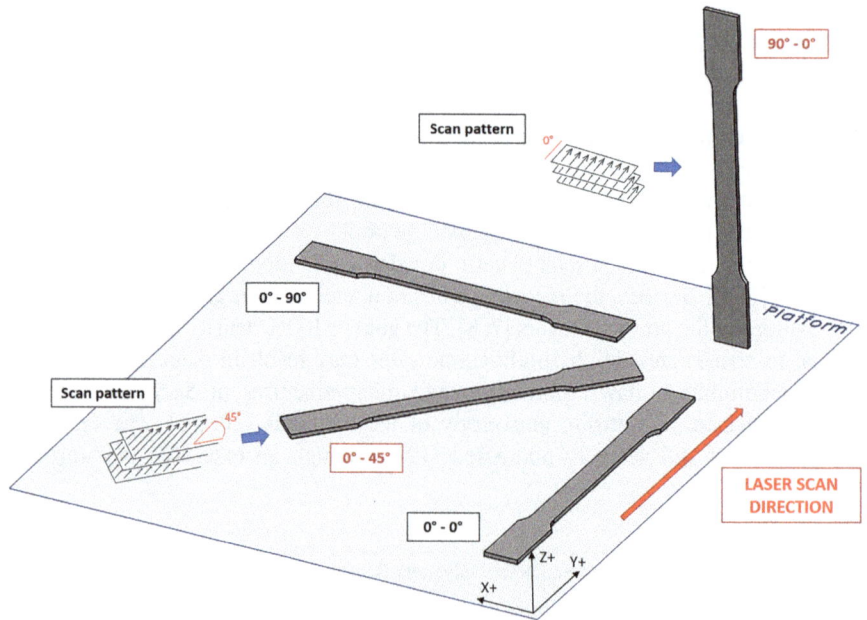

Fig. 12.1 Dog-bone specimens are produced with different orientations with respect to the platform and laser scan direction

Fig. 12.2 Experimental setup

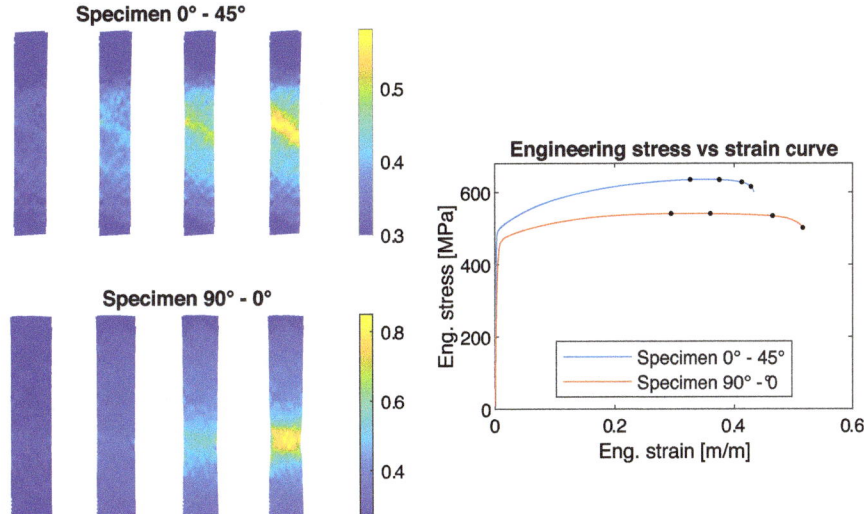

Fig. 12.3 Longitudinal strain map obtained by DIC in the two specimens in the post-necking region. The maps are taken at different steps of the tests, highlighted by the black dot in the stress vs. strain curve shown on the right

Digital Image Correlation was performed using a customized software [18] that exploits global DIC. In particular, a mesh of 42×10 elements was used, where each element is a squared subset of 16×10 pixels. A particular type of rubber-like paint was used to ensure that the speckle pattern does not break before the final facture, accommodating with the large deformation of the specimens.

12.3   Analysis

The result of the DIC analysis as well as the engineering stress vs. strain curves for the two experiments are illustrated in Fig. 12.3. On the left, the maps show the distribution of the longitudinal strain component ε_y in the post-necking region for both specimens. The first map represents the strain distribution at the onset of necking, where the stress-strain state is still mainly uniaxial. The last map, instead, is the strain map obtained just before fracture, where the necking is fully developed and the strain localization is clearly visible. The corresponding points in the stress-strain curves are highlighted by black dots in the graph on the right.

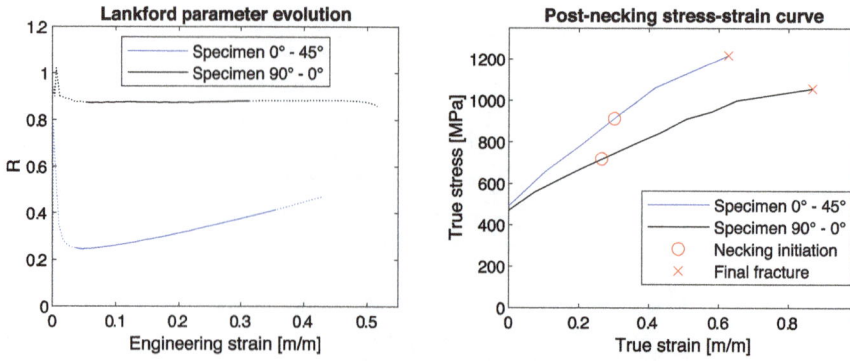

Fig. 12.4 Anisotropy evaluation of the two specimen and true-stress true strain curve in the post-necking regime, identified using the VFM

Although the metal powder material used for the AM process is the same, the different paths of the SLM machine adopted to manufacture the specimens produce materials with completely different properties. This is clearly visible from Fig. 12.3: the initial yield stress is rather similar, ranging between 470 and 490 MPa, but then the strain hardening of the specimen formed at 0°–45° is more pronounced leading to a maximum stress of 635 MPa, around 20% higher than the one of the 90°–0° specimen, i.e., 540 MPa.

The VFM was used to characterize the post-necking behavior in terms of true-stress true-strain curve up to the final fracture. Each specimen was considered as a separate material; the Hill48 model with normal anisotropy was used as yielding function while the stress strain curve was derived as a linear piece-wise curve [19]. The first graph of Fig. 12.4 shows the evolution of the Lanford parameter R, i.e., the ratio of the transversal strain to the through-thickness strain, during the two tests as a function of the engineering strain. The solid lines represent the zones were the computed R-value is significant, i.e., when the stress state is uniaxial, excluding the initial elastic and yielding phase and the final necking zone, where the stress state is multiaxial.

The 0°–45° specimen shows a high anisotropy, with R ranging from 0.3 to 0.4. This value is much lower than expected for this material and it is clearly due to the AM process. Moreover, the R-value is not constant during the test. To simplify the analysis, in this work an average constant value of $R = 0.35$ was used in the VFM algorithm to evaluate the post-necking curve. On the contrary, the 90°–45° specimen presents a rather standard anisotropic behavior, with a stable R-value during plastic deformation. In this case a value of $R = 0.86$ was achieved, that falls within a typical range for these materials.

The true-stress vs. true-strain curves obtained with VFM are presented in the second graph of Fig. 12.4; in the same graph the points of necking initiation and final fracture are also highlighted. As discussed for anisotropy, a very different post-necking behavior is observed in the two cases. For the 0°–45° specimen, the necking starts at $\varepsilon_{necking} = 0.3$ with a final accumulated plastic strain at fracture $\varepsilon_{fracture} = 0.63$. For the 90°–0° specimen, instead, the necking starts at $\varepsilon_{necking} = 0.26$ with a final accumulated plastic strain at fracture $\varepsilon_{fracture} = 0.87$. Thus, it turns out that a flat specimen formed in a plane parallel to the machine platform, like the 0°–45° one, has higher resistance and lower ductility, with a reduced final strain at fracture. A specimen formed perpendicularly to the machine platform, instead, presents a higher ductility and most of the plastic deformation occurs in the post-necking regime.

Interestingly, the onset of necking occurs first in the 90°–0° specimen, although it has a larger ductility. The shape of the necking is also quite different, as illustrated in Fig. 12.3. For the 0°–45° specimen, the necking is rather sharp with a narrow strain localization almost inclined at 45%. On the other hand, the necking in the 90°–0° specimen is diffuse, with the strain localization perpendicular to the specimen axis. It must be pointed out here that the final strain at fracture obtained with 2D DIC is the superficial one; in order to evaluate the fracture strain inside the specimen further analyses could be conducted (see for instance [20, 21]) that, however, are beyond the scope of this paper.

12.4   Conclusion

The paper presents the identification of the post-necking behavior in specimen produced with AM, in particular using the SLM method on steel powder. Two flat specimens were produced, the first parallel to the plane of the machine basement and, accordingly, to the powder deposition layer; the second perpendicular to the basement. Tensile tests were conducted on the specimens and the strain maps up to finale fracture measured using DIC. Finally, the anisotropic behavior and the true-stress true-strain curves in the post-necking region were evaluated using the VFM. The main outcomes of this paper are:

- The path of the laser scan during the melting process has a large impact in the post-necking behavior of the flat specimens formed with the SLM method.
- The AM process parameters affect the anisotropy of the material, the strain hardening, and the final fracture/damage evolution.
- Flat specimens formed in a plane parallel to the powder layer have better mechanical properties in terms of strain hardening and onset of necking, but a reduced ductility.
- A very low R-value is observed in the $0°–45°$ specimen.

This work demonstrates that, in order to validate the crashworthiness and the safety of AM products, it is necessary to carefully investigate the mechanical behavior up to the final fracture. The post-necking behavior is highly influenced by the process parameters indeed. Starting from this preliminary analysis, in future works, extensive tests will be conducted looking at different orientation and scan patterns.

References

1. Srivastava, M., Rathee, S., Maheshwari, S., Kundra, T.K.: Additive Manufacturing. CRC Press, New York (2019)
2. Milewski, J.O.: Additive manufacturing of metals. In: Springer Series in Materials Science. Springer, New York (2017)
3. Khorasani, A., Gibson, I., Veetil, J.K., Ghasemi, A.H.: A review of technological improvements in laser-based powder bed fusion of metal printers. Int. J. Adv. Manuf. Technol. **108**, 191–209 (2020)
4. Brandt, M. (ed.): Laser additive manufacturing: materials, design, technologies, and applications. Woodhead Publishing, Cambridge (2016)
5. Ghouse, S., Babu, S., Nai, K., Hooper, P.A., Jeffers, J.R.T.: The influence of laser parameters, scanning strategies and material on the fatigue strength of a stochastic porous structure. Addit. Manuf. **22**, 290–301 (2018)
6. Brika, S.E., Letenneur, M., Dion, C.A., Brailovski, V.: Influence of particle morphology and size distribution on the powder flowability and laser powder bed fusion manufacturability of Ti-6Al-4V alloy. Addit. Manuf. **31**, 100929 (2020)
7. Reijonen, J., Revuelta, A., Riipinen, T., Ruusuvuori, K., Puukko, P.: On the effect of shielding gas flow on porosity and melt pool geometry in laser powder bed fusion additive manufacturing. Addit. Manuf. **32**, 101030 (2020)
8. Snow, Z., Nassar, A.R., Reutzel, E.W.: Invited review article: review of the formation and impact of flaws in powder bed fusion additive manufacturing. Addit. Manuf. **36**, 101457 (2020)
9. Hooper, P.A.: Melt pool temperature and cooling rates in laser powder bed fusion. Addit. Manuf. **22**, 548–559 (2018)
10. Ronneberg, T., Davies, C.M., Hooper, P.A.: Revealing relationships between porosity, microstructure and mechanical properties of laser powder bed fusion 316L stainless steel through heat treatment. Mater. Des. **189**, 108481 (2020)
11. Im, Y.-D., Kim, K.-H., Jung, K.-H., Lee, Y.-K., Song, K.-H.: Anisotropic mechanical behavior of additive manufactured AISI 316L steel. Metall Mat Trans. **50**, 2014–2021 (2019)
12. Nalli, F., Cortese, L., Concli, F.: Ductile damage assessment of Ti6Al4V, 17-4PH and AlSi10Mg for additive manufacturing. Eng. Fract. Mech. **241**, 107395 (2021)
13. Cortese, L., Nalli, F., Rossi, M.: A nonlinear model for ductile damage accumulation under multiaxial non-proportional loading conditions. Int. J. Plast. **85**, 77–92 (2016)
14. Attilio, L., Barlat, F., Pierron, F., Marek, A., Rossi, M.: Inverse identification strategies for the characterization of transformation based anisotropic plasticity models with the non-linear VFM. Int. J. Mech. Sci. **173**, 105422 (2020)
15. Sasso, M., Mancini, E., Chiappini, G., Sarasini, F., Tirillò, J.: Application of DIC to static and dynamic testing of agglomerated cork material experimental mechanics. Exp Mech. **58**(7), 1017–1033 (2018)
16. Stazi, F., Tittarelli, F., Saltarelli, F., Chiappini, G., Morini, A., Cerri, G., Lenci, S.: Carbon nanofibers in polyurethane foams: experimental evaluation of thermo-hygrometric and mechanical performance. Polym. Test. **67**, 234–245 (2018)
17. Badaloni, M., Rossi, M., Chiappini, G., Lava, P., Debruyne, D.: Impact of experimental uncertainties on the identification of mechanical material properties using DIC. Exp. Mech. **55**(8), 1411–1426 (2015)
18. Amodio, D., Broggiato, G.B., Campana, F., Newaz, G.M.: Digital speckle correlation for strain measurement by image analysis. Exp. Mech. **43**(4), 396–402 (2003)
19. Rossi, M., Lattanzi, A., Barlat, F.: A general linear method to evaluate the hardening behaviour of metals at large strain with full-field measurements. Strain. **54**(3), e12265 (2018)
20. Rossi, M., Chiappini, G., Sasso, M.: Characterization of aluminum alloys using a 3D full field measurement. Soc Exp Mech. **1**, 93–99 (2010)
21. Rossi, M., Cortese, L., Genovese, K., Lattanzi, A., Nalli, F., Pierron, F.: Evaluation of volume deformation from surface DIC measurement. Exp. Mech. **58**(7), 1181–1194 (2018)
22. Utzeri, M., Sasso, M., Chiappini, G., Lenci, S.: Nonlinear vibrations of a composite beam in large displacements: analytical, numerical, and experimental approaches. J Comput Nonlin Dyn. **16**, 2 (2020)

Chapter 13
Heat Source Reconstruction Applied to Fatigue Characterization Under Varying Stress Amplitude

C. Douellou, A. Gravier, X. Balandraud, and E. Duc

Abstract Fatigue damage is associated with heat release leading to material self-heating. The present study deals with the measurement of mechanical dissipation from temperature measurements by infrared (IR) thermography during a fatigue test with varying stress amplitude. The experiment was performed in two steps on an additively manufactured steel specimen. First, specific acquisition conditions of the thermal response were used to remove the cyclic fluctuation due to thermoelastic coupling. Second, heat source reconstruction was applied to evaluate the calorific origin of the self-heating during the test. A simplified version of the heat diffusion equation was used assuming a uniform distribution of mechanical dissipation within the specimen. The relationship between stress amplitude and mechanical dissipation was identified without the need for a steady thermal regime at constant load amplitude.

Keywords Infrared thermography · Intrinsic dissipation · Self-heating · Fatigue limit · Steel · Additive manufacturing

13.1 Introduction

Under cyclic mechanical loading at a given frequency, a material heats up more or less depending on the mean and amplitude of stress applied. This is due to a release of *mechanical dissipation* (also called *intrinsic dissipation*) accompanying the fatigue damage. The first observations of self-heating during a fatigue test were made by Stromeyer at the beginning of the twentieth century [1]. In the 1990s, infrared (IR) thermography was used to study the relationship between the self-heating of the material and its fatigue limit [2–5]. The main idea is to distinguish two regimes in the thermal response, i.e., below and above the fatigue limit of the material. In practice, a material specimen is subjected to several sequences of mechanical cycles at constant stress amplitude. Each sequence is usually applied for a certain time in order to achieve a stationary thermal oscillation of the material specimen. During the test, the constant stress amplitude is gradually increased until the specimen breaks. A return time to thermal equilibrium can be respected between the sequences so that the "reference" initial thermal state can be properly redefined as the test progresses. Indeed, temperature of the specimen environment may vary due to changes in ambient air temperature and the grips of the testing machine (in particular when using a hydraulic actuator, as is often the case for fatigue tests). The total duration of the test depends on the number of stress amplitudes applied. Throughout the test, an IR camera is dedicated to measure the temperature of the specimen. The extraction of the thermal signature of the fatigue damage for each stress amplitude is then performed. Classically, two quantities can be identified from the *mean* temperature evolution: the rate of mean temperature rise at the beginning of the cyclic loading sequence; or the mean temperature change in steady thermal regime (with respect to the initial temperature before the cyclic loading was applied). The thermal signature is then plotted as a function of the stress amplitude. The treatment of this type of graph consists in distinguishing different trends at "low" and "high" stress amplitudes. At the low stress amplitudes, self-heating remains low and increases slightly with the stress amplitude. For materials featuring a fatigue limit (below which the service life is a priori infinite) such as steels, the self-heating increases significantly with the stress amplitude. This significant self-heating is considered to be related to fatigue damage leading to ultimate failure of the specimen. The boundary between the two thermal regimes is considered to be the fatigue limit of the tested material. Graphical methods (also referred to as "deterministic" in Ref. [6]) have been proposed to distinguish these two regimes in the fatigue thermal signature as a function of the stress amplitude:

C. Douellou · A. Gravier · X. Balandraud (✉) · E. Duc
Institut Pascal, Clermont Auvergne INP, CNRS, Université Clermont Auvergne, Clermont-Ferrand, France
e-mail: xavier.balandraud@sigma-clermont.fr

© The Society for Experimental Mechanics, Inc. 2022
S. L. B. Kramer et al. (eds.), *Thermomechanics & Infrared Imaging, Inverse Problem Methodologies, Mechanics of Additive & Advanced Manufactured Materials, and Advancements in Optical Methods & Digital Image Correlation, Volume 4*, Conference Proceedings of the Society for Experimental Mechanics Series, https://doi.org/10.1007/978-3-030-86745-4_13

- Intersection of the linear regression at "high" stress amplitudes with the abscissa axis;
- Intersection of the linear regressions at "high" and "low" stress amplitudes.

In both cases, the abscissa of the intersection is considered as the boundary between the two self-heating regimes, i.e., the value of the fatigue limit of the material. Other criteria have been compared for identifying the separation between the two thermal regimes using more robust protocols than the intersection of linear trends [6]. Specific approaches have been discussed when temperature stabilization is not achieved during the loading at constant stress amplitude [7]. Also, works have been carried out on the understanding of the physical mechanisms related to self-heating in fatigue, namely the progressive appearance of micro-plasticity during cyclic loading [8–11]. In particular, the authors have identified the microstructural mechanisms appearing under cyclic loading at low stress amplitude (high cycle fatigue) and have shown that the corresponding self-heating can be described by a quadratic function as a function of the stress amplitude. Assessment of fatigue limits has been performed on various materials such as different types of steels [12–15], magnesium alloys [16, 17], titanium alloys [18], as well as composite materials [19, 20].

In previous works, we have proposed a model of mechanical dissipation as a function of stress amplitude, as well as different criteria for identifying the fatigue limit of additively manufactured steels [21–23]. More precisely, we used a "heat source reconstruction" technique to determine the mechanical dissipation (in $W.m^{-3}$) at the origin of the temperature change measured by IR camera. The data treatment is based on the heat diffusion equation [24–26]. We used a macroscopic approach (called "0D approach") considering the *mean* temperature variation over the gauge zone of the specimen [27]. Finally, we applied specific acquisition conditions of the temperature maps to extract only the part of the self-heating associated with mechanical dissipation. In practice, the temperatures recorded by the camera were averaged over an integer number of mechanical cycles, which allows canceling the cyclic fluctuation associated with thermoelastic coupling. In conclusion, the heat source reconstruction technique allowed us to measure a calorific data (in $W.m^{-3}$) associated with the fatigue response of the material (namely the mechanical dissipation), independently of the heat exchanges by conduction with the test machine jaws and by convection with the ambient air. The approach was applied to compare the fatigue response of maraging steels elaborated by additive manufacturing, namely by Selective Laser Melting (SLM) process [21]. It was possible to quickly distinguish the fatigue performance of specimens differing in their manufacturing parameters. In the present work, heat source reconstruction was carried out to measure the mechanical dissipation under cyclic loading with continuously varying stress amplitude, i.e., without imposing a waiting time at each stress amplitude value.

13.2 Experimental Setup

The experimental setup is presented in Fig. 13.1. A schematic view is given in Fig. 13.1a, where it can be seen that the plane specimen featured two "reference" zones on both sides of the mechanically loaded zone. The role of these reference zones is to track the thermal changes in the environment during the test, see details in Ref. [28]. In particular, they enabled us to follow the slight changes in the temperature of the two grips of the testing machine. A $\pm$ 15 kN MTS machine was used to apply the mechanical loading. It is equipped with a hydraulic actuator, which creates a slow temperature increase of the moving grip when the machine is functioning due to its working oil. The cross-sectional area of the mechanically tested area of the

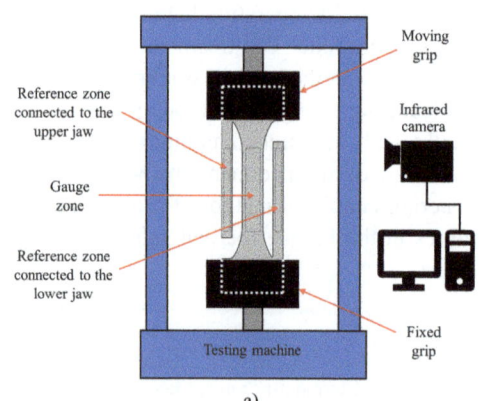
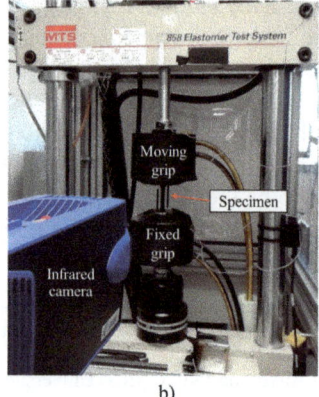

a)   b)

Fig. 13.1 Experimental setup: a) schematic view, b) picture of the experiment

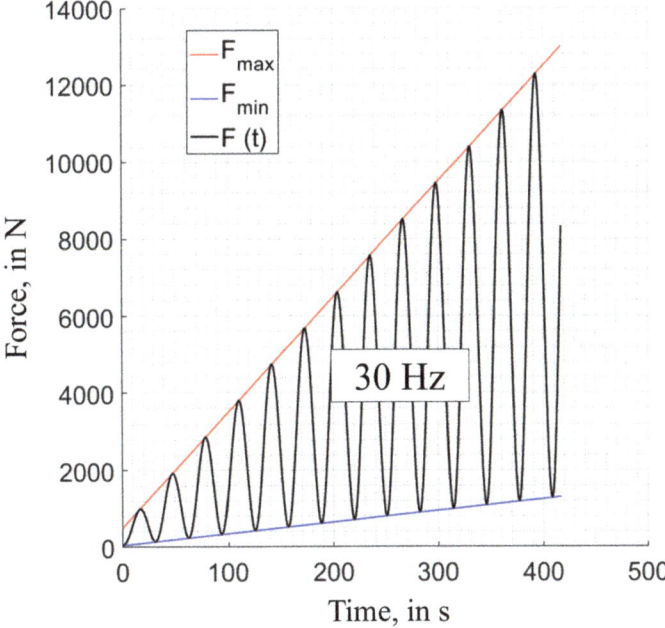

Fig. 13.2 Schematic representation of the mechanical loading

specimen was equal to 10 mm^2. Loading consisted of sinusoidal force-controlled cycles with continuously increasing amplitudes until specimen failure; see the schematic representation in Fig. 13.2. The loading frequency was set to 30 Hz. The force ratio between the minimum and maximum forces was kept constant at 0.1. The maximum stress was increased by 0.1 MPa per cycle. This particular loading enables a fast characterization procedure because heat source reconstruction allows an assessment of mechanical dissipation without the need for a steady thermal regime at constant load amplitude.

Figure 13.1-b shows a picture of the experiment. The maraging steel specimen was painted in black to increase the thermal emissivity. A FLIR camera was utilized to measure the temperature on the surface of the reference zones as well as on the mechanically loaded zone of the specimen. The base acquisition frequency of the camera was set to 147 Hz. However, the recording frequency was equal to 1 Hz by applying an averaging operation every 147 images. This enabled us to remove the fluctuation due to the thermoelastic coupling. Temperatures were also captured for 10 s just before the beginning of the cyclic loading. This enabled us to define the reference thermal state of the specimen and thus to calculate the temperature change during the cyclic loading. The variation of the macroscopic (0D) temperature change θ of the mechanically loaded zone was established taken into account the drift in the temperature of the specimen's close environment thanks to the two reference zones (see Ref. [28] for details).

13.3 Thermomechanical Background and Heat Source Reconstruction

Thermomechanical couplings lead to heat release or absorption by a material when it is submitted to a mechanical loading. In the following, the term "heat source" will refer to the heat power density (in W.m^{-3}) released or absorbed by the material itself due to a change in its mechanical state. The total heat source s is composed of a *mechanical dissipation* part (d_1) and a *thermoelastic coupling* part ($s_{\text{th} - \text{el}}$):

$$s = d_1 + s_{\text{th}-\text{el}} \tag{13.1}$$

Irreversible mechanical phenomena are accompanied by a production of mechanical dissipation: $d_1 > 0$. In the case of a fatigue loading, this heat release is the calorific origin of the material self-heating. On the contrary, thermoelastic couplings lead to heat production or absorption as a function of the sign of the stress rate. For isotropic materials featuring an isentropic thermoelastic coupling such as metals, the corresponding heat source writes

$$s_{\text{th}-\text{el}} = -\alpha\, T\, \frac{\mathrm{d}\sigma}{\mathrm{d}t} \tag{13.2}$$

where α is the coefficient of thermal expansion (in K^{-1}), T the temperature (in K), and σ the sum of the principal stresses (in Pa). As T is in Kelvin, it can be considered equal to the initial temperature T_0 in the case of thermal changes limited to some degrees during the test. For uniaxial tensile states as in the present experiment, σ is the axial stress.

Heat sources can be reconstructed from the knowledge of temperatures measured on the specimen surface by IR thermography. The processing is based on the heat diffusion equation. When the heat sources s are spatially homogeneous within the specimen (in particular for homogeneous stress and strain states in the gauge zone), the reconstruction can be done using the *average* temperature change θ over the whole gauge zone [17]. It is then assumed that the heat exchanges by conduction with jaws of the testing machine and by convection with ambient air are both proportional to θ. The simplified heat eq. (13.3) provides the relationship between the heat source $s(t)$ and the mean temperature change $\theta(t)$ along the test:

$$s(t) = \rho C\left(\frac{\mathrm{d}\theta(t)}{\mathrm{d}t} + \frac{\theta(t)}{\tau}\right) \tag{13.3}$$

where ρ is the material density (in $kg.m^{-3}$), C the specific heat (in $J.kg^{-1}.K^{-1}$), and τ a time constant (in s) characterizing the global heat exchanges of the specimen with its environment (ambient air and jaws of the testing machine).

Thermoelastic coupling is a "strong" coupling. During a fatigue loading, $s_{\text{th}-\text{el}}(t)$ is much higher in magnitude than $d_1(t)$. However, measuring mechanical dissipation from the thermal response of the material is possible for a cyclic loading at constant stress amplitude. Indeed, Eq. (13.2) shows that the heat originating from the thermoelastic coupling is null over a thermodynamical cycle. So, mechanical dissipation can be directly obtained from Eq. (13.3) using the average temperature change θ over integer numbers of mechanical cycles. In the case of a cyclic loading with continuously varying stress amplitude, the thermoelastic coupling heat does not vanish, but its value can be simply evaluated from the time integral of Eq. (13.2) over the considered integer number of cycles.

13.4   Results

Figure 13.3 shows the variation of the mean temperature change θ during the fatigue test featuring a maximum stress increasing by 0.1 MPa per cycle. Let us recall that the temporal resolution of θ is equal to 1 s as a consequence of the averaging operation of the thermal signal by the IR camera. It can be seen in the graph that the temperature change is first slightly negative, as a consequence of the thermoelastic coupling effect with low mechanical dissipation. The curve is quite flat at the beginning than significantly increases. Specimen failure occurred at about 415 MPa, for a temperature change of about 5.4 K. Figure 13.4 shows the results of the heat source reconstruction. The red solid curve corresponds to the mechanical dissipation d_1 measured for the present test. The stress resolution (on the x-axis) is equal to 3 MPa. This value is a consequence of the temporal resolution of the temperature measurement at 1 Hz (stress increase of 1 MPa per second) and the calculation of the term $\mathrm{d}\theta/\mathrm{d}t$ in Eq. (13.3) by centered finite difference scheme (which penalizes the temporal resolution by a factor of three). Black crosses in Fig. 13.4 correspond to mechanical dissipation values measured for cyclic loading sequences at different *constant* stress amplitudes on the same maraging steel in Ref. [21]. It can be observed that the two layouts are very similar, validating the approach with continuously varying stress amplitude.

13.5   Conclusion

Fatigue damage is associated with a release of mechanical dissipation. In the present study, we succeeded in measuring this calorific quantity from the thermal data measured by an IR camera during a fatigue test with varying stress amplitude. Application was done on an additively manufactured maraging steel. Thanks to specific acquisition conditions of the temperatures at the surface of the specimen, it was possible to remove the cyclic fluctuation due to thermoelastic coupling. Heat source reconstruction was then applied to calculate the mechanical dissipation using a simplified version of the heat diffusion equation. The link between mechanical dissipation and stress amplitude was established without the need for a steady thermal regime at constant load amplitude, which opens up the perspectives of much faster fatigue characterization procedures.

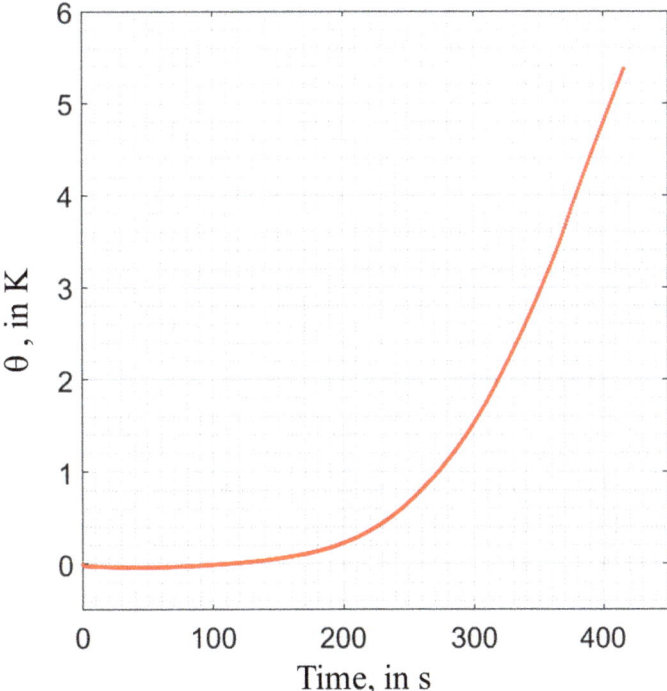

Fig. 13.3 Variation of the mean temperature change θ during the cyclic loading with continuously varying stress amplitude (0.1 MPa per mechanical cycle)

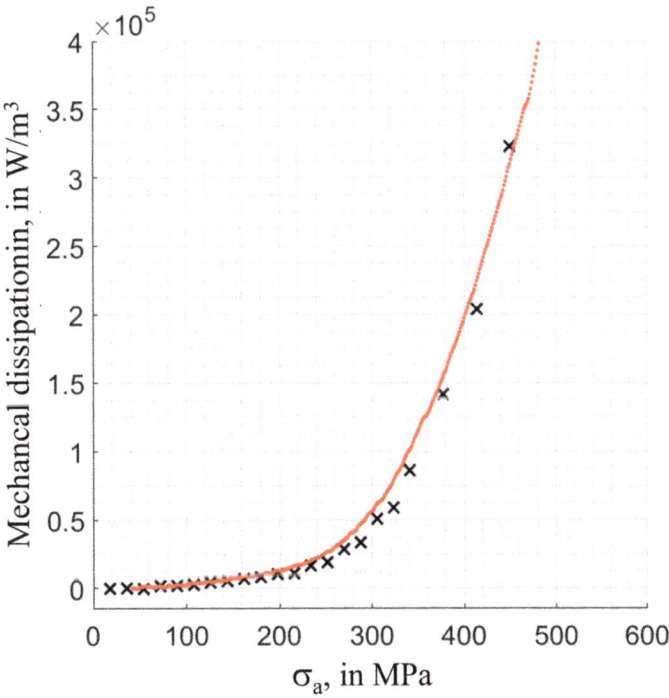

Fig. 13.4 Results of the heat source reconstruction: in red, mechanical dissipation d_1 during the cyclic loading with continuously varying stress amplitude (0.1 MPa per mechanical cycle); in black, mechanical dissipation d_1 for sequences of mechanical cycles at various constant stress amplitudes

Acknowledgments The authors acknowledge the Région Auvergne-Rhônes-Alpes for the support in this study (Project: IRICE Fabrication additive, number: 18 009727 01-59941, operation: P088O005).

References

1. Stromeyer, C.E.: The determination of fatigue limits under alternating stress conditions. Proc Royal Soc London. **90**, 411–425 (1914)
2. Luong, M.P.: Infrared thermographic scanning of fatigue in metals. Nucl. Eng. Des. **158**, 363–376 (1995)
3. Luong, M.P.: Fatigue limit evaluation of metals using an infrared thermographic technique. Mech. Mater. **28**, 155–163 (1998)
4. Geraci A.L., La Rosa G., Risitano A., Grech M.: Determination of the fatigue limit of an austempered ductile iron using thermal infrared imagery, In: Fedosov E.A. (ed.) Proceedings of SPIE. Digital Photogrammetry and Remote Sensing '95, Moscow, Russia, June 25–30, 1995, Vol. 2646, pp. 306–317 (1995)
5. La Rosa, G., Risitano, A.: Thermographic methodology for rapid determination of the fatigue limit of materials and mechanical components. Int. J. Fatigue. **22**, 65–73 (2000)
6. Huang, J., Pastor, M.L., Garnier, C., Gong, X.: Rapid evaluation of fatigue limit on thermographic data analysis. Int. J. Fatigue. **104**, 293–301 (2017)
7. de Finis, R., Palumbo, D., da Silva, M.M., Galietti, U.: Is the temperature plateau of a self-heating test a robust parameter to investigate the fatigue limit of steels with thermography? Fatigue Fract. Eng. Mater. Struct. **41**, 917–934 (2018)
8. Doudard, C., Calloch, S., Hild, F., Cugy, P., Galtier, A.: Identification of the scatter in high cycle fatigue from temperature measurements. Comptes Rendus Mécanique. **332**, 795–801 (2004)
9. Doudard, C., Calloch, S., Cugy, P., Galtier, A., Hild, F.: A probabilistic two-scale model for high cycle fatigue life predictions. Fatigue Fract. Eng. Mater. Struct. **28**, 279–288 (2005)
10. Munier, R., Doudard, C., Calloch, S., Weber, B.: Determination of high cycle fatigue properties of a wide range of steel sheet grades from self-heating measurements. Int. J. Fatigue. **63**, 46–61 (2014)
11. Munier, R., Doudard, C., Calloch, S., Weber, B.: Identification of the micro-plasticity mechanisms at the origin of self-heating under cyclic loading with low stress amplitude. Int. J. Fatigue. **103**, 122–135 (2017)
12. De Finis, R., Palumbo, D., Ancona, F., Galietti, U.: Fatigue limit evaluation of various martensitic stainless steels with new robust thermographic data analysis. Int. J. Fatigue. **74**, 88–96 (2015)
13. Cao, Y.F., Moumni, Z., Zhu, J.H., Zhang, Y.H., You, Y.J., Zhang, W.H.: Comparative investigation of the fatigue limit of additive-manufactured and rolled 316 steel based on self-heating approach. Eng. Fract. Mech. **223**, 106746 (2020)
14. Shiozawa, D., Inagawa, T., Washio, T., Sakagami, T.: Fatigue limit estimation of stainless steels with new dissipated energy data analysis. Procedia Struct Integr. **2**, 2091–2096 (2016)
15. Stankovicova, Z., Dekys, V., Uhricik, M., Novak, P., Strnadel, B.: Fatigue limit estimation using IR camera. In: Vasko, M., et al. (eds.) XXII Slovak-Polish Scientific Conference on Machine Modelling and Simulations 2017 (MMS 2017), p. 05021. EDP Sciences, Les Ulis (2018). https://doi.org/10.1051/matecconf/201815705021
16. Guo, S.F., Liu, X.S., Zhang, H.X., Yan, Z.F., Zhang, Z.D., Fang, H.Y.: Thermographic study of AZ31B magnesium alloy under cyclic loading: temperature evolution analysis and fatigue limit estimation. Materials. **13**, 5209 (2020)
17. Guo, S.F., Liu, X.S., Zhang, H.X., Yan, Z.F., Fang, H.Y.: Fatigue limit evaluation of AZ31B magnesium alloy based on temperature distribution analysis. Metals. **10**, 1331 (2020)
18. Akai A., Shiozawa D., Sakagami T., Fatigue limit estimation of titanium alloy Ti-6Al-4V with infrared thermography, In: Bison P., Burleigh D. (eds) Proceedings of SPIE. Thermosense: Thermal Infrared Applications XXXIX, Anaheim, CA, April 10–13, 2017, Vol. 10214, 102141J (2017)
19. De Finis, R., Palumbo, D., Galietti, U.: Fatigue damage analysis of composite materials using thermography-based techniques. Procedia Struct Integr. **18**, 781–791 (2019)
20. Palumbo, D., De Finis, R., Demelio, P.G., Galietti, U.: A new rapid thermographic method to assess the fatigue limit in GFRP composites. Compos Part B-Eng. **103**, 60–67 (2016)
21. Douellou, C., Balandraud, X., Duc, E., Verquin, B., Lefebvre, F., Sar, F.: Rapid characterization of the fatigue limit of additive-manufactured maraging steels using infrared measurements. Addit. Manuf. **35**, 101310 (2020)
22. Douellou, C., Balandraud, X., Duc, E.: Fatigue characterization of 3D-printed maraging steel by infrared thermography. In: Kramer, S., et al. (eds.) Mechanics of Additive and Advanced Manufacturing, Vol 8. Conference Proceedings of the Society for Experimental Mechanics Series, pp. 5–9. Springer, Cham (2019). https://doi.org/10.1007/978-3-319-95083-9_2
23. Douellou, C., Balandraud, X., Duc, E., Verquin, B., Lefebvre, F., Sar, F.: Fast fatigue characterization by infrared thermography for additive manufacturing. Procedia Struct Integr. **19**, 90–100 (2019)
24. Chrysochoos, A., Peyroux, R.: Experimental analysis and numerical simulation of thermomechanical couplings in solid materials. Revue Générale de Thermique. **37**, 582–606 (1998)
25. Boulanger, T., Chrysochoos, A., Mabru, C., Galtier, A.: Calorimetric analysis of dissipative and thermoelastic effects associated with the fatigue behavior of steels. Int. J. Fatigue. **26**, 221–229 (2004)
26. Chrysochoos, A., Louche, H.: An infrared image processing to analyse the calorific effects accompanying strain localization. Int. J. Eng. Sci. **38**, 1759–1788 (2000)
27. Jongchansitto, P., Douellou, C., Preechawuttipong, I., Balandraud, X.: Comparison between 0D and 1D approaches for mechanical dissipation measurement during fatigue tests. Strain. **55**, e12307 (2019)
28. Delpueyo, D., Balandraud, X., Grediac, M., Stanciu, S., Cimpoesu, N.: A specific device for enhanced measurement of mechanical dissipation in specimens subjected to long-term tensile tests in fatigue. Strain. **54**, e12252 (2018)

Chapter 14
Analysis of the Thermomechanical Response of a Rubber-like Granular Material

K. Jongchansitto, P. Jongchansitto, I. Preechawuttipong, J. -B. Le Cam, F. Blanchet, M. Grédiac, and X. Balandraud

Abstract Infrared thermography was used to study the thermal signature of a rubber-like granular material subjected to cyclic confined compression. The discrete medium consisted of cylinders placed in parallel. A fluctuation of the temperature at the same frequency as the mechanical loading was observed, as well as a global self-heating as the cycles progressed. This can be associated with the thermoelastic coupling effect and the mechanical dissipation effect, respectively. The thermoelastic coupling effect is visible in all contact zones between the particles, which can be explained by the stress concentrations that occur there. A strong mechanical dissipation effect occurs at specific contacts, which can be explained by the high friction between certain particles.

Keywords Granular material · Rubber-like material · Infrared thermography · Thermoelastic coupling · Intrinsic dissipation

14.1 Introduction

Granular materials are omnipresent around us, in many industrial fields and in many natural phenomena. Rice and sugar are used in the food industry and in our daily lives. Sand and rocks are commonly used in civil engineering for the construction of buildings or streets. Granular flows are involved in avalanches and landslides. In the context of engineering, granular materials can be defined as a set of solid particles whose macroscopic mechanical behavior is governed by interparticle forces, i.e., contact forces between particles. Granular media are generally made up of grains having various distributions in terms of size, shape, and base material. This leads to a variety of mechanical phenomena that are not yet clearly understood. Numerical simulations have been widely used to study the influence of parameters such as particle shape, density, polydispersity, elasticity, and friction [1]. Several experimental techniques are also available. For measurements in the volume, we can mention X-ray tomography [2]. For surface measurements, particle image velocimetry and digital image correlation are widely used [3–9]. Stress fields can be obtained by photoelasticimetry measurement using particles made of birefringent materials [10–12]. Mechanoluminescent materials have also been used to visualize the intensity of interparticle contacts [13]. In most of the studies available in the literature, the particles are rigid or hard. Contact stiffness is then the main parameter. A granular system consisting of soft matter is more complex to analyze. Recently, the compaction of highly deformable particle assemblies has been performed numerically in Ref. [14], opening the possibility of comparison with experimental results for soft granular materials.

Infrared (IR) thermography has also been used to measure the thermal signature of the mechanical response of soils and sands [15–17]. Some studies have mainly focused on Schneebeli materials (two-dimensional discrete media made with

K. Jongchansitto
Department of Mechanical Engineering, Faculty of Engineering, Chiang Mai University, Chiang Mai, Thailand

Institut Pascal, Clermont Auvergne INP, CNRS, Université Clermont Auvergne, Clermont-Ferrand, France

P. Jongchansitto · I. Preechawuttipong
Department of Mechanical Engineering, Faculty of Engineering, Chiang Mai University, Chiang Mai, Thailand

J. -B. Le Cam · F. Blanchet
Institut de Physique, Université de Rennes 1, Rennes Cedex, France

M. Grédiac · X. Balandraud (✉)
Institut Pascal, Clermont Auvergne INP, CNRS, Université Clermont Auvergne, Clermont-Ferrand, France
e-mail: xavier.balandraud@sigma-clermont.fr

© The Society for Experimental Mechanics, Inc. 2022
S. L. B. Kramer et al. (eds.), *Thermomechanics & Infrared Imaging, Inverse Problem Methodologies, Mechanics of Additive & Advanced Manufactured Materials, and Advancements in Optical Methods & Digital Image Correlation, Volume 4*, Conference Proceedings of the Society for Experimental Mechanics Series, https://doi.org/10.1007/978-3-030-86745-4_14

cylinders) in order to analyze the contact zones between the particles, i.e., the stress concentration zones [18–22]. In the present study, IR thermography was used to analyze the thermal response of Schneebeli media made of thermoplastic polyurethane (TPU). Temperature changes were measured with an IR camera during cyclic confined compression. The objective was to distinguish the thermoelastic coupling effect (associated with reversible mechanical behavior) and the mechanical dissipation effect (associated with irreversible mechanical behavior).

14.2   Experimental Preparation

The experimental setup is presented in Fig. 14.1. A granular sample was prepared by randomly placing cylinders with elliptical cross-section into a rectangular metallic frame: see the schematic view in Fig. 14.1a. The particles were made of TPU manufactured by using an in-house device. Each particle was molded at ambient temperature in a vacuum chamber to avoid bubble formation. One side of the elliptical face of each cylinder was polished to obtain a smooth surface perpendicular to the longitudinal axis. To maximize the thermal emissivity of the surfaces observed by the camera (and thus reduce reflections in the IR range), a matt black paint was sprayed on the particle cross-sections as well as on the frame containing the granular system. Black curtains and cardboards were also placed around the device (not visible in Fig. 14.1b) to minimize parasitic reflections.

A cyclic triangular compressive loading was applied on the top of the granular system by using a Zwick/Roell ZMART PRO testing machine through a pusher (see Fig. 14.1b). The minimum and maximum forces were set to −1200 N and − 12,000 N, respectively. The loading was force-controlled with a force rate equal to 900 N/s, leading to cycle duration of 24 s. Note that some preliminary mechanical cycles were applied in order to compact the granular system. Next, the minimum force was kept constant for about 10 min to ensure that a steady thermal equilibrium of the system was reached before starting the cyclic loading.

A Cedip Jade III-MWIR camera was utilized to capture the temperature fields on the surface of each elliptical particle during the mechanical loading at a recording frequency of 10 Hz. In practice, the base acquisition frequency of the camera was set to 100 Hz with a real-time averaging every ten images in order to improve the thermal measurement resolution. The resulting standard deviation of the thermal noise (measurement resolution) was lower than 0.01 °C thanks to this real-time average operation.

Temperature maps were also captured during 5 seconds just before the beginning of the cyclic loading in order to define the reference temperature field. Figure 14.2a shows the field of temperature differences at the beginning of the mechanical test with respect to this reference thermal state in order to reveal the spatial noise during the test. The corresponding distribution of the values within the particles (i.e., excluding the voids between the particles) is displayed in Fig. 14.2b. Values lower than about −0.04 °C are due to pixels near the boundary of the particles. The standard deviation of the distribution within the [−0.04 °C; 0.04 °C] range is equal to 0.012 °C.

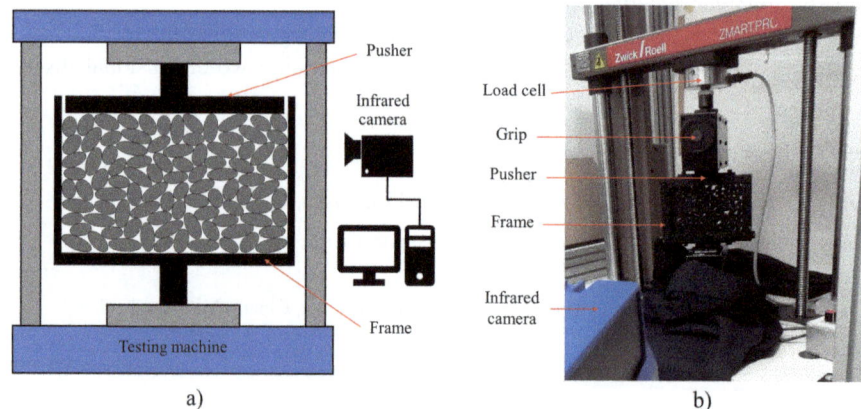

a)   b)

Fig. 14.1 Experimental setup: a) schematic view, b) picture of the experiment

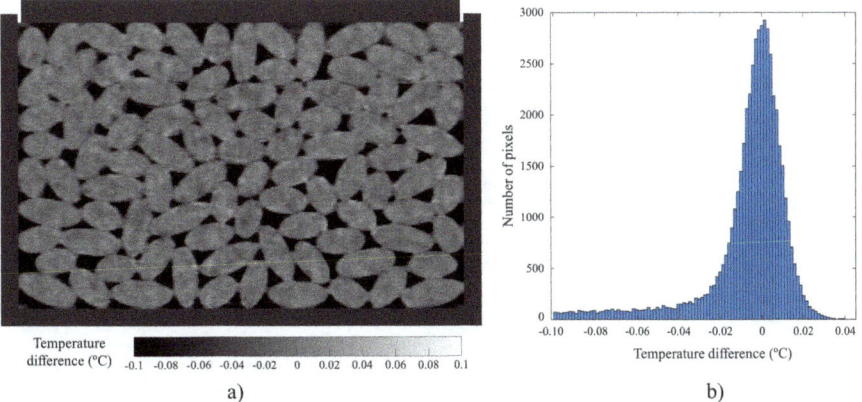

Fig. 14.2 Thermal noise: (**a**) temperature differences at the beginning of the cyclic loading with respect to the reference temperature field, (**b**) corresponding distribution for the pixels within the particles

14.3   Thermomechanical Background

Thermomechanical couplings in materials can be separated into two different parts, namely a part associated to *reversible* processes (elasticity) and a part associated to *irreversible* processes (e.g., plasticity, viscosity, fatigue damage, friction):

– The first case is called *thermoelastic coupling*. Two types of coupling between temperature and strain must be considered according to the physics associated with elasticity: *isentropic* coupling governed by thermal expansion or compression; and *entropic* coupling governed by the degree of disorder. In rubber-like materials, entropic coupling becomes dominant over isentropic coupling as soon as the level of deformation is greater than 10% [23].
– In the second case, the associated calorific quantity d_1 (in W/m^3) is called *mechanical dissipation* or *intrinsic dissipation*. It is worth remembering that it is different from the thermal dissipation d_2, the two types of dissipation being involved in the Clausius-Duhem inequality $d_1 + d_2 > 0$.

Mechanical dissipation can be revealed through the temperature change over integer numbers of mechanical cycles. Indeed, heat quantities due to reversible mechanical phenomena are null over a thermodynamical cycle. More precisely, the heat quantities upon loading and unloading are equal in magnitude but opposite in sign. On the contrary, the mechanical dissipation is positive whatever the loading evolution: $d_1 > 0$. Thermoelastic coupling is a "strong" coupling. Indeed, the corresponding heat power density is much higher in magnitude than the mechanical dissipation in most cases. Extracting mechanical dissipation effect from the global thermal response of the material is in general possible for a cyclic loading, for which thermoelastic coupling effect vanishes cyclically. Moreover, the heat produced by irreversible phenomena accumulates over the cycles (because d_1 is always positive), which helps to experimentally identify this quantity. In the present experiment, thermoelastic coupling was evidenced by considering half of the first mechanical cycle (assuming that the mechanical dissipation is low over this duration) whereas mechanical dissipation was highlighted by considering ten cycles.

14.4   Analysis of the Reversible Mechanical Phenomena

Figure 14.3 shows the temperature change field at the half of the first mechanical cycle. "Hot" zones are clearly evidenced at each contact between the particles. They are also visible at the contacts with the metallic frame. They are logically due to the stress concentrations resulting mainly a priori from the normal interparticle forces. The largest temperature increase is observed in the upper left corner of the granular material. Tangential forces may be also involved in this zone, as revealed by the analysis of the thermal signature of the irreversible mechanical phenomena in the next section.

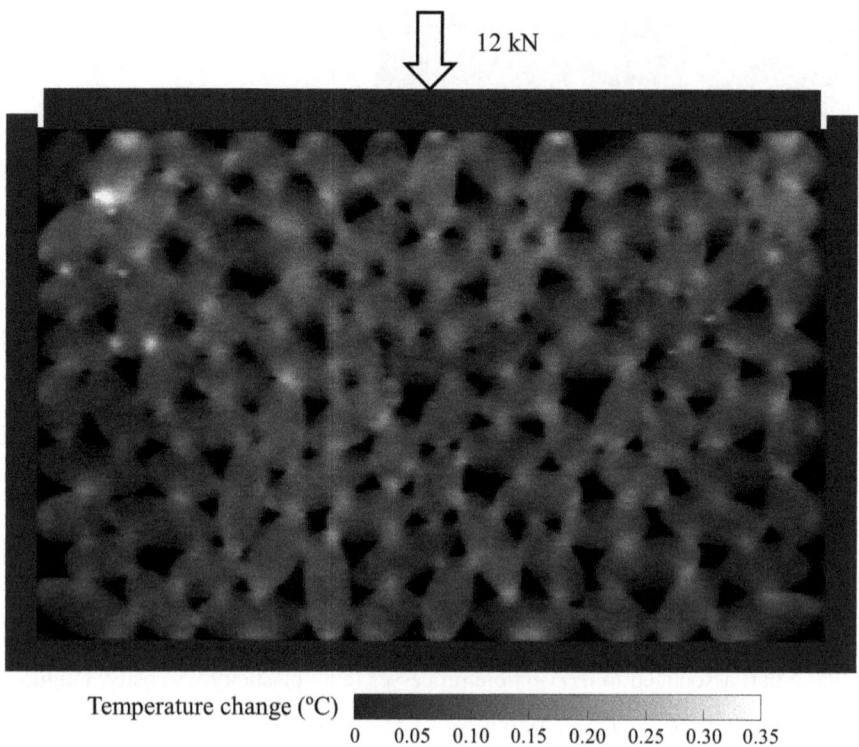

12 kN

Temperature change (°C)

0   0.05 0.10 0.15 0.20 0.25 0.30 0.35

Fig. 14.3 Field of temperature change at the half of the first mechanical cycle, evidencing the thermoelastic coupling effect at the interparticle contacts in the granular system under confined compression

14.5   Analysis of the Irreversible Mechanical Phenomena

Figure 14.4 presents the field of temperature change at the end of the tenth mechanical cycle. Based on the assumption that the heat quantity due to the thermoelastic coupling is null over the preceding successive mechanical cycles, this map displays a thermal signature of the mechanical dissipation effect, i.e., of the irreversible mechanical phenomena. In granular media, this latter quantity can be mainly associated to friction at the contacts. However, viscosity and damage in the volume can be also involved, especially in the zones subjected to high stress levels. Figure 14.4 clearly shows that the mechanical dissipation effect is not present at all the contacts, which strongly differs from the case of the thermoelastic coupling effect (compare with Fig. 14.3). "Hot" zones are now mainly visible only in the upper two corners of the granular system. This can be explained by the greater movement of the particles in the upper two corners, potentially leading to greater friction. It can also be observed that the size of the "hot" zones is quite large, which can be explained by the heat diffusion over the ten cycles.

Finally, Fig. 14.5 shows the variation in time of the temperature at a given interparticle contact located in the upper left part of the granular material: see point A in the schematic view. As expected, the temperature oscillates at the same frequency as the mechanical loading. It can be seen that the mean temperature change first increases over the cycles and then stabilizes from nearly the seventh cycle: see the red dashed curve. This increase can be attributed to the monotonous production of mechanical dissipation over the cycles. The stabilization results from the equilibrium between the heat produced by the mechanical irreversible phenomena at point A and the heat diffused to the environment of point A.

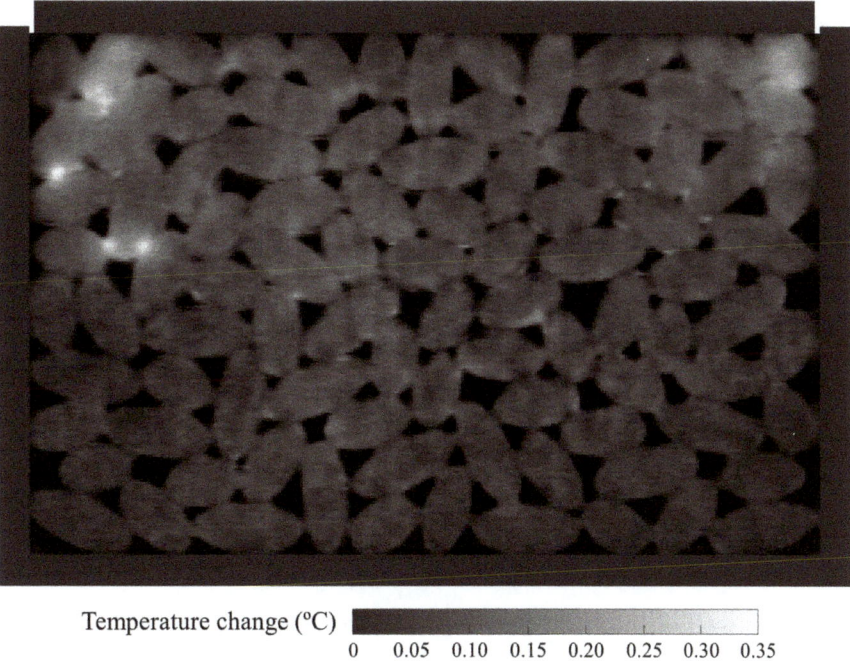

Fig. 14.4 Field of temperature change at the end of the tenth mechanical cycles, evidencing the mechanical dissipation effect especially in the two top corners of the granular material

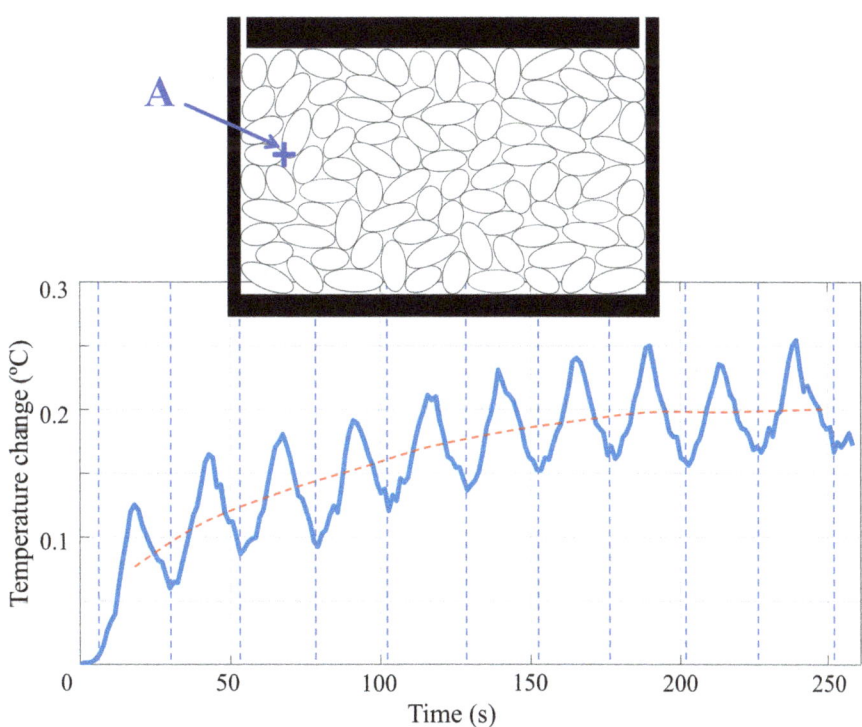

Fig. 14.5 Temperature variation at the interparticle contact point A along the cyclic loading (blue curve). The red dashed curve corresponds to the global self-heating

14.6 Conclusion

IR thermography is in principle applicable to analyze thermomechanical phenomena of any types of material. However, it remains difficult to apply to cohesionless granular materials, which explains the small numbers of studies published in the literature (see Refs [15–22]). The objective of this study was to analyze the thermomechanical response of a two-dimensional granular system made from TPU cylinders with elliptical cross-section. A cyclic compression loading was applied to a granular system while an IR camera measured the resulting temperature changes. The thermoelastic coupling effect and the mechanical dissipation effect were distinguished in the thermal response. The thermoelastic coupling effect is clearly visible in the contact zones between the particles, which can be explained by the stress concentrations occurring there. The mechanical dissipation effect manifests itself in specific contacts, which can be explained by the high friction between certain particles.

Acknowledgments K. Jongchansitto would like to acknowledge the National Research Council of Thailand (NRCT) through the Royal Golden Jubilee Ph.D. program (grant no. PHD/0100/2561) for the support during this research.

References

1. Radjai, F., Roux, J.N., Daouadji, A.: Modeling granular materials: century-long research across scales. J. Eng. Mech. **143**, 04017002 (2017)
2. Khalili, M.H., Brisard, S., Bornert, M., Aimedieu, P., Pereira, J.M., Roux, J.N.: Discrete digital projections correlation: a reconstruction-free method to quantify local kinematics in granular media by X-ray tomography. Exp. Mech. **57**, 819–830 (2017)
3. Slominski, C., Niedostatkiewicz, M., Tejchman, J.: Application of particle image velocimetry (PIV) for deformation measurement during granular silo flow. Powder Technol. **173**, 1–18 (2007)
4. Hall, S.A., Wood, D.M., Ibraim, E., Viggiani, G.: Localised deformation patterning in 2D granular materials revealed by digital image correlation. Granul. Matter. **12**, 1–14 (2010)
5. Richefeu, V., Combe, G., Viggiani, G.: An experimental assessment of displacement fluctuations in a 2D granular material subjected to shear. Geotech Lett. **2**, 113–118 (2012)
6. Karanjgaokar, N., Ravichandran, G.: Experimental inference of inter-particle forces in granular systems using digital image correlation. In: Jin, H., Sciammarella, C., Yoshida, S., et al. (eds.) Advancement of Optical Methods in Experimental Mechanics, pp. 379–385. Springer, Berlin (2014)
7. Karanjgaokar, N., Ravichandran, G.: Study of energy contributions in granular materials during impact. In: Song, B., Lamberson, L., Casem, D., et al. (eds.) Dynamic Behavior of Materials, p. 199. Springer, New York (2015)
8. Hurley, R., Marteau, E., Ravichandran, G., Andrade, J.E.: Extracting inter-particle forces in opaque granular materials: beyond photoelasticity. J. Mech. Phys. Solids. **63**, 154–166 (2014)
9. Hurley, R.C., Lim, K.W., Ravichandran, G., Andrade, J.E.: Dynamic inter-particle force inference in granular materials: method and application. Exp. Mech. **56**, 217–229 (2016)
10. Shukla, A., Damania, C.: Experimental investigation of wave velocity and dynamic contact stresses in an assembly of disks. Exp. Mech. **27**, 268–281 (1987)
11. Roessig, K.M., Foster, J.C., Bardenhagen, S.G.: Dynamic stress chain formation in a two-dimensional particle bed. Exp. Mech. **42**, 329–337 (2002)
12. Mirbagheri, S.A., Ceniceros, E., Jabbarzadeh, M., McCormick, Z., Fu, H.C.: Sensitively photoelastic biocompatible gelatin spheres for investigation of locomotion in granular media. Exp. Mech. **55**, 427–438 (2015)
13. Jongchansitto, P., Boyer, D., Preechawuttipong, I., Balandraud, X.: Using mechanoluminescent materials to visualize interparticle contact intensity in granular media. Exp. Mech. **60**, 51–64 (2020)
14. Cantor, D., Cardenas-Barrantes, M., Preechawuttipong, I., Renouf, M., Azema, E.: Compaction model for highly deformable particle assemblies. Phys. Rev. Lett. **124**, 208003 (2020)
15. Luong, M.P.: Characteristic threshold and infrared vibrothermography of sand. Geotech. Test. J. **9**, 80–86 (1986)
16. Luong, M.P.: Infrared Thermography of the Dissipative Behaviour of Sand, pp. 199–202. A.A. Balkema Publishers, Leiden, Netherlands (2001)
17. Luong, M.P.: Introducing infrared thermography in soil dynamics. Infrared Phys Technol. **49**, 306–311 (2007)
18. Jongchansitto, P., Balandraud, X., Grédiac, M., Beitone, C., Preechawuttipong, I.: Using infrared thermography to study hydrostatic stress networks in granular materials. Soft Matter. **10**, 8603–8607 (2014)
19. Chaiamarit, C., Balandraud, X., Preechawuttipong, I., Grédiac, M.: Stress network analysis of 2D non-cohesive polydisperse granular materials using infrared thermography. Exp. Mech. **39**, 761–769 (2015)
20. Jongchansitto, P., Balandraud, X., Preechawuttipong, I., Le Cam, J.-B., Garnier, P.: Thermoelastic couplings and interparticle friction evidenced by infrared thermography in granular materials. Exp. Mech. **58**, 1469–1478 (2018)
21. Jongchansitto, P., Preechawuttipong, I., Balandraud, X., Grédiac, M.: Numerical investigation of the influence of particle size and particle number ratios on texture and force transmission in binary granular composites. Powder Technol. **308**, 324–333 (2017)
22. Jongchansitto, P., Yachai, T., Preechawuttipong, I., Boufayed, R., Balandraud, X.: Concept of mechanocaloric granular material made from shape memory alloy. Energy. **219**, 119656 (2021)
23. Balandraud, X., Le Cam, J.B.: Some specific features and consequences of the thermal response of rubber under cyclic mechanical loading. Arch. Appl. Mech. **84**, 773–788 (2014)

Chapter 15
Which Pattern for a Low Pattern-Induced Bias?

Frédéric Sur, Benoît Blaysat, and Michel Grédiac

Abstract The first objective of this presentation is to show that it is possible to explain the cause of the pattern-induced bias (PIB) observed in displacement fields obtained by local DIC. A model is presented for this purpose. It gathers the different errors made when retrieving this displacement by minimizing the optical residual over subsets. It is shown that PIB predicted with this model and its counterpart observed with displacement fields obtained with DIC are in good agreement. When DIC is applied on periodic patterns like checkerboards instead of random speckles, it is observed that PIB becomes negligible. Such regular patterns are however not well suited for DIC. Hence it is recalled how to process such images by minimizing the optical residual in the Fourier domain instead of the spatial one. PIB is assessed in this case, and it is also observed that PIB is negligible in displacement maps obtained with such regular patterns processed by minimizing the optical residual in the Fourier domain.

Keywords Checkerboard · Digital image correlation · Pattern-induced bias · Localized spectrum analysis · Uncertainty quantification

15.1 Statement of the Problem

Quantifying the uncertainty in measurements obtained with DIC is a topical issue in the experimental mechanics community, certainly because this is an essential prerequisite before standardizing these techniques. There is a rather wide literature concerning random errors in DIC, in particular sensor noise propagation to the final displacement maps, [1] for instance. The systematic error caused by interpolation performed to express the deformed image in the coordinate system of the reference image is also the aim of many papers, [2] for instance. Another important systematic error is the so-called matching bias, which manifests itself by a "damping" of the actual details in displacement or strain maps. It can be regarded as the effect of a convolution of the true displacement with a low-pass Savitzky-Golay (SG) filter [3]. This latter result is quite satisfactory on average, but it does not account for the effect of the pattern involved in the determination of the displacement field. Indeed, the interplay between matching functions, actual displacement field, image gray level distribution, and its gradients causes spurious fluctuations to appear around the displacement field resulting from the aforementioned convolution. These spurious fluctuations can be regarded as a bias recently named pattern-induced bias [4, 5] (PIB). It induces a random distribution of "blobs," which impair the displacement maps.

F. Sur (✉)
Laboratoire Lorrain de Recherche en Informatique et ses Applications, Université de Lorraine, CNRS, INRIA projet Tangram, Campus Scientifique, Vandoeuvre-lès-Nancy Cedex, France
e-mail: frederic.sur@loria.fr

B. Blaysat · M. Grédiac
Institut Pascal, Université Clermont Auvergne, Clermont Auvergne INP, CNRS, Clermont-Ferrand, France

S. L. B. Kramer et al. (eds.), *Thermomechanics & Infrared Imaging, Inverse Problem Methodologies, Mechanics of Additive & Advanced Manufactured Materials, and Advancements in Optical Methods & Digital Image Correlation, Volume 4*, Conference Proceedings of the Society for Experimental Mechanics Series, https://doi.org/10.1007/978-3-030-86745-4_15

15.2 Model

PIB can be predicted as a function of various parameters, namely the speckle itself, its gradient, the difference between actual displacement field and its approximation by subset shape functions being the most influencing ones [6]. DIC consists in minimizing over subsets the following residual with respect to the set parameters $\lambda_j j = 1..N$, gathered in a vector denoted by Λ:

$$\text{SSD}(\Lambda) = \sum_{x_i \in \Omega_x} \left(\mathcal{I}(x_i) - \widetilde{\mathcal{I}'}\left(x_i + \sum_{j=1}^{N} \lambda_j \phi_j(x_i)\right)\right)^2 \tag{15.1}$$

where $\mathcal{I}$ is the reference image, $\widetilde{\mathcal{I}'}$ the deformed image, and Φ_j, $j = 1..N$, the shape functions. λ_j, $j = 1..N$, are then used to describe the actual displacement fields within the subsets. Assuming that the images are not affected by noise, it has been demonstrated in [6] that at convergence of the minimization of the SSD function, vector Λ gathering the parameters returned by DIC was equal to the following quantity.

$$\Lambda = \Lambda^{\mathbf{u}} + \left((L^{\mathbf{u}})^T L^{\mathbf{u}}\right)^{-1}(L^{\mathbf{u}})^T \mathbf{E} + \left((L^{\mathbf{u}})^T L^{\mathbf{u}}\right)^{-1}(L^{\mathbf{u}})^T \mathbf{D}\mathcal{I}' \tag{15.2}$$

where the coefficient in row i and column j of matrice $L^{\mathbf{u}}$ is

$$L_{i,j}^{\mathbf{u}} = \left\langle \nabla \mathcal{I}'(x_i + \mathbf{u}(x_i)), \phi_j(x_i) \right\rangle \tag{15.3}$$

This coefficient represents the projection of the image gradient on the shape functions. In Eq. (15.2), the first of the three terms is the set of parameters $\Lambda^{\mathbf{u}}$ that would be obtained by projecting the true displacement on the shape functions, but this true displacement remains generally unknown in practice. This vector is also the convolution of the true displacement by a SG filter [3]. $\mathbf{E}$ contained in the second term represents the projection of the deformed image gradient on the residual of the decomposition of the displacement over the basis of shape functions, and $\mathbf{D}\mathcal{I}'$ in the third term is the subpixel interpolation error. The case of noisy images is also addressed in [6] but it is not discussed here for the sake of simplicity.

The model proposed here enables us to analyze the displacement field returned by DIC, including the effect of PIB. We choose a reference displacement containing a gently decreasing spatial frequency [7, 8] to illustrate PIB. The ability of the model above to predict this phenomenon is also illustrated, see Fig. 15.1 where various displacement fields are shown. PIB is indeed the difference between the displacement measured by DIC and the reference displacement field convolved by the SG filter corresponding to the subset used in DIC in terms of width and order of the matching function. The cross-sections of four displacement fields are depicted in this figure (bottom right). It can be seen that the red curve obtained with the model correctly

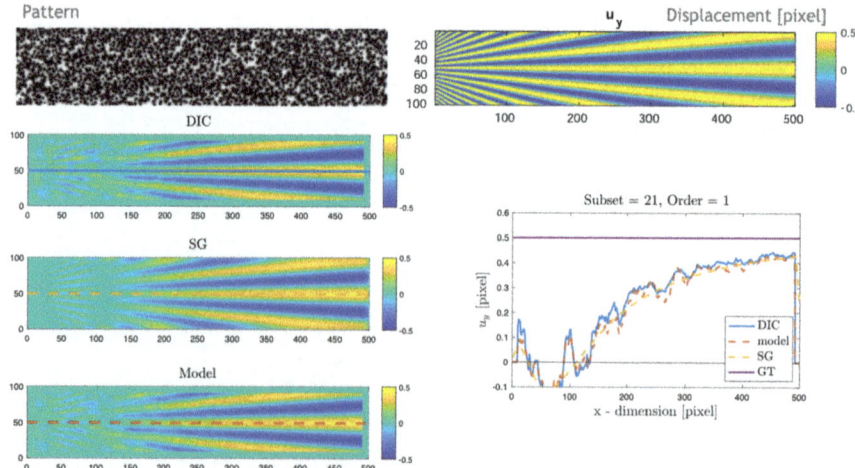

Fig. 15.1 Top left: artificial speckle deformed through the reference displacement field shown at the top right. The displacement field returned by DIC (subset size: 21 pixels, first-order matching functions), the reference displacement field convolved by the corresponding Savitzky-Golay filter, and the displacement field retrieved by using by Eq. (15.2) are successively given. Bottom right: cross-section of the different displacement fields along the symmetry axis of these three maps

predicts the main fluctuations of the blue curve obtained with DIC. The difference with the reference displacement convolved by the SG filter is significant, in particular at the left-hand side of the curve, thus for the highest spatial frequencies involved in the displacement field.

15.3 Case of Pariodic Patterns Like Checkerboard

In [5, 9], it is shown that periodic patterns lead to a PIB which is negligible compared to the one found with random speckles. The problem is however that such periodic patterns are not well suited to DIC because of their periodicity. It is however worth mentioning that the minimization of the SSD function, usually carried out iteratively in the spatial domain with DIC, can advantageously be switched to the Fourier domain when periodic patterns are considered [9]. In this case, the minimization is quasi-direct, which also dramatically speeds up the calculations [10]. It can also be shown that the PIB discussed above becomes negligible when this minimization is performed in the Fourier domain with periodic patterns like checkerboards. Various examples will be given and discussed during the presentation to illustrate this result. The main limitation of this model is that the image gradient is involved in Eq. (15.2), and correctly assessing this quantity is difficult with real patterns used in experimental mechanics, the signal contained in numerical images being sampled.

References

1. Blaysat, B., Grédiac, M., Sur, F.: On the propagation of camera sensor noise to displacement maps obtained by DIC. Exp. Mech. **56**(6), 919–944 (2016)
2. Schreier, H.W., Braasch, J.R., Sutton, M.: Systematic errors in digital image correlation caused by intensity interpolation. Optim. Eng. **39**(11), 2915–2921 (2000)
3. Schreier, H.W., Sutton, M.A.: Systematic errors in digital image correlation due to undermatched subset shape functions. Exp. Mech. **42**(3), 303–310 (2002)
4. Lehoucq, R.B., Reu, P.L., Turner, D.Z.: The effect of the ill-posed problem on quantitative error assessment in digital image correlation. Exp Mech. **2017**, 1–13 (2017)
5. Fayad, S., Seidl, D.T., Reu, P.L.: Spatial DIC errors due to pattern-induced bias and grey level discretization. Exp. Mech. **60**(2), 249–263 (2020)
6. Sur, F., Blaysat, B., Grédiac, M.: On biases in displacement estimation for image registration, with a focus on photomechanics. J Math Imag Vision. **63**, 777–806 (2021)
7. Grédiac, M., Sur, F.: Effect of sensor noise on the resolution and spatial resolution of the displacement and strain maps obtained with the grid method. Strain. **50**(1), 1–27 (2014)
8. 2D Challenge 2.0 discussion document. https://sem.org/dicchallenge
9. Grédiac, M., Blaysat, B., Sur, F.: A critical comparison of some metrological parameters characterizing local digital image correlation and grid method. Exp. Mech. **57**(6), 871–903 (2017)
10. Grédiac, M., Blaysat, B., Sur, F.: On the optimal pattern for displacement field measurement: random speckle and DIC, or checkerboard and LSA? Exp. Mech. **60**(4), 509–534 (2020)

9 783030 867478